Ford Escort
RS Cosworth
& World Rally Car

Also from Veloce Publishing:

Speedpro Series
4-cylinder Engine – How To Blueprint & Build A Short Block For High Performance (Hammill)
Alfa Romeo DOHC High-performance Manual (Kartalamakis)
Alfa Romeo V6 Engine High-performance Manual (Kartalamakis)
BMC 998cc A-series Engine – How To Power Tune (Hammill)
1275cc A-series High-performance Manual (Hammill)
Camshafts – How To Choose & Time Them For Maximum Power (Hammill)
Competition Car Datalogging Manual, The (Templeman)
Cylinder Heads – How To Build, Modify & Power Tune Updated & Revised Edition (Burgess & Gollan)
Distributor-type Ignition Systems – How To Build & Power Tune (Hammill)
Fast Road Car – How To Plan And Build Revised & Updated Colour New Edition (Stapleton)
Ford SOHC 'Pinto' & Sierra Cosworth DOHC Engines – How To Power Tune Updated & Enlarged Edition (Hammill)
Ford V8 – How To Power Tune Small Block Engines (Hammill)
Harley-Davidson Evolution Engines – How To Build & Power Tune (Hammill)
Holley Carburetors – How To Build & Power Tune Revised & Updated Edition (Hammill)
Jaguar XK Engines – How To Power Tune Revised & Updated Colour Edition (Hammill)
MG Midget & Austin-Healey Sprite – How To Power Tune Updated & Revised Edition (Stapleton)
MGB 4-cylinder Engine – How To Power Tune (Burgess)
MGB V8 Power – How To Give Your, Third Colour Edition (Williams)
MGB, MGC & MGB V8 – How To Improve (Williams)
Mini Engines – How To Power Tune On A Small Budget Colour Edition (Hammill)
Motorcycle-engined Racing Car – How To Build (Pashley)
Motorsport – Getting Started in (Collins)
Nitrous Oxide High-performance Manual, The (Langfield)
Rover V8 Engines – How To Power Tune (Hammill)
Sportscar/kitcar Suspension & Brakes – How To Build & Modify Revised & Updated 3rd Edition (Hammill)
SU Carburettor High-performance Manual (Hammill)
Suzuki 4x4 – How To Modify For Serious Off-road Action (Richardson)
Tiger Avon Sportscar – How To Build Your Own Updated & Revised 2nd Edition (Dudley)
TR2, 3 & TR4 – How To Improve (Williams)
TR5, 250 & TR6 – How To Improve (Williams)
TR7 & TR8 – How To Improve (Williams)
V8 Engine – How To Build A Short Block For High Performance (Hammill)
Volkswagen Beetle Suspension, Brakes & Chassis – How To Modify For High Performance (Hale)
Volkswagen Bus Suspension, Brakes & Chassis – How To Modify For High Performance (Hale)
Weber DCOE, & Dellorto DHLA Carburetors – How To Build & Power Tune 3rd Edition (Hammill)

Those Were The Days ... Series
Alpine Trials & Rallies 1910-1973 (Pfundner)
Austerity Motoring (Bobbitt)
Brighton National Speed Trials (Gardiner)
British Police Cars (Walker)
British Woodies (Peck)
Dune Buggy Phenomenon (Hale)
Dune Buggy Phenomenon Volume 2 (Hale)
Hot Rod & Stock Car Racing in Britain In The 1980s (Neil)
MG's Abingdon Factory (Moylan)
Motor Racing At Brands Hatch In The Seventies (Parker)
Motor Racing At Crystal Palace (Collins)
Motor Racing At Goodwood In The Sixties (Gardiner)
Motor Racing At Nassau In The 1950s & 1960s (O'Neil)
Motor Racing At Oulton Park In The 1960s (Mcfadyen)
Motor Racing At Oulton Park In The 1970s (Mcfadyen)
Three Wheelers (Bobbitt)

Enthusiast's Restoration Manual Series
Citroën 2CV, How To Restore (Porter)
Classic Car Bodywork, How To Restore (Thaddeus)
Classic Car Electrics (Thaddeus)
Classic Cars, How To Paint (Thaddeus)
Reliant Regal, How To Restore (Payne)
Triumph TR2/3/3A, How To Restore (Williams)
Triumph TR4/4A, How To Restore (Williams)
Triumph TR5/250 & 6, How To Restore (Williams)
Triumph TR7/8, How To Restore (Williams)
Volkswagen Beetle, How To Restore (Tyler)
VW Bay Window Bus (Paxton)
Yamaha FS1-E, How To Restore (Watts)

Essential Buyer's Guide Series
Alfa GT (Booker)
Alfa Romeo Spider Giulia (Booker & Talbott)
BMW GS (Henshaw)
BSA Bantam (Henshaw)
BSA Twins (Henshaw)
Citroën 2CV (Paxton)
Citroën ID & DS (Heilig)
Fiat 500 & 600 (Bobbitt)
Jaguar E-type 3.8 & 4.2-litre (Crespin)
Jaguar E-type V12 5.3-litre (Crespin)
Jaguar/Daimler XJ6, XJ12 & Sovereign (Crespin)
Jaguar XJ-S (Crespin)
MGB & MGB GT (Williams)
Mercedes-Benz 280SL-560DSL Roadsters (Bass)
Mercedes-Benz 'Pagoda' 230SL, 250SL & 280SL Roadsters & Coupés (Bass)
Morris Minor & 1000 (Newell)
Porsche 928 (Hemmings)
Rolls-Royce Silver Shadow & Bentley T-Series (Bobbitt)
Subaru Impreza (Hobbs)
Triumph Bonneville (Henshaw)
Triumph TR6 (Williams)
VW Beetle (Cservenka & Copping)
VW Bus (Cservenka & Copping)

Auto-Graphics Series
Fiat-based Abarths (Sparrow)
Jaguar MKI & II Saloons (Sparrow)
Lambretta Li Series Scooters (Sparrow)

Rally Giants Series
Audi Quattro (Robson)
Austin Healey 100-6 & 3000 (Robson)
Fiat 131 Abarth (Robson)
Ford Escort MkI (Robson)
Ford Escort RS Cosworth & World Rally Car (Robson)
Ford Escort RS1800 (Robson)
Lancia Stratos (Robson)
Peugeot 205 T16 (Robson)
Subaru Impreza (Robson)

General
1½-litre GP Racing 1961-1965 (Whitelock)
AC Two-litre Saloons & Buckland Sportscars (Archibald)
Alfa Romeo Giulia Coupé GT & GTA (Tipler)
Alfa Romeo Montreal – The Essential Companion (Taylor)
Alfa Tipo 33 (McDonough & Collins)
Alpine & Renault – The Development Of The Revolutionary Turbo F1 Car 1968 to 1979 (Smith)
Anatomy Of The Works Minis (Moylan)
Armstrong-Siddeley (Smith)
Autodrome (Collins & Ireland)
Automotive A-Z, Lane's Dictionary Of Automotive Terms (Lane)
Automotive Mascots (Kay & Springate)
Bahamas Speed Weeks, The (O'Neil)
Bentley Continental, Corniche And Azure (Bennett)
Bentley MkVI, Rolls-Royce Silver Wraith, Dawn & Cloud/Bentley R & S-Series (Nutland)
BMC Competitions Department Secrets (Turner, Chambers, Browning)
BMW 5-Series (Cranswick)
BMW Z-Cars (Taylor)
Britains Farm Model Balers & Combines 1967 to 2007 (Pullen)
British 250cc Racing Motorcycles (Pereira)
British Cars, The Complete Catalogue Of, 1895-1975 (Culshaw & Horrobin)
BRM – A Mechanic's Tale (Salmon)
BRM V16 (Ludvigsen)
BSA Bantam Bible, The (Henshaw)
Bugatti Type 40 (Price)
Bugatti 46/50 Updated Edition (Price & Arbey)
Bugatti T44 & T49 (Price & Arbey)
Bugatti 57 2nd Edition (Price)
Caravans, The Illustrated History 1919-1959 (Jenkinson)
Caravans, The Illustrated History From 1960 (Jenkinson)
Carrera Panamericana, La (Tipler)
Chrysler 300 – America's Most Powerful Car 2nd Edition (Ackerson)
Chrysler PT Cruiser (Ackerson)
Citroën DS (Bobbitt)
Cliff Allison – From The Fells To Ferrari (Gauld)
Cobra – The Real Thing! (Legate)
Cortina – Ford's Bestseller (Robson)
Coventry Climax Racing Engines (Hammill)
Daimler SP250 New Edition (Long)
Datsun Fairlady Roadster To 280ZX – The Z-Car Story (Long)
Dino – The V6 Ferrari (Long)
Dodge Charger – Enduring Thunder (Ackerson)
Dodge Dynamite! (Grist)
Donington (Boddy)
Draw & Paint Cars – How To (Gardiner)
Drive On The Wild Side, A – 20 Extreme Driving Adventures From Around The World (Weaver)
Ducati 750 Bible, The (Falloon)
Ducati 860, 900 And Mille Bible, The (Falloon)
Dune Buggy, Building A – The Essential Manual (Shakespeare)
Dune Buggy Files (Hale)
Dune Buggy Handbook (Hale)
Edward Turner: The Man Behind The Motorcycles (Clew)
Fiat & Abarth 124 Spider & Coupé (Tipler)
Fiat & Abarth 500 & 600 2nd Edition (Bobbitt)
Fiats, Great Small (Ward)
Fine Art Of The Motorcycle Engine, The (Peirce)
Ford F100/F150 Pick-up 1948-1996 (Ackerson)
Ford F150 Pick-up 1997-2005 (Ackerson)
Ford GT – Then, And Now (Streather)
Ford GT40 (Legate)
Ford In Miniature (Olson)
Ford Model Y (Roberts)
Ford Thunderbird From 1954, The Book Of The (Long)
Forza Minardi! (Vigar)
Funky Mopeds (Skelton)
Gentleman Jack (Gauld)
GM In Miniature (Olson)
GT – The World's Best GT Cars 1953-73 (Dawson)
Hillclimbing & Sprinting – The Essential Manual (Short & Wilkinson)
Honda NSX (Long)
Jaguar, The Rise Of (Price)
Jaguar XJ-S (Long)
Jeep CJ (Ackerson)
Jeep Wrangler (Ackerson)
Karmann-Ghia Coupé & Convertible (Bobbitt)
Lamborghini Miura Bible, The (Sackey)
Lambretta Bible, The (Davies)
Lancia 037 (Collins)
Lancia Delta HF Integrale (Blaettel & Wagner)
Land Rover, The Half-ton Military (Cook)
Laverda Twins & Triples Bible 1968-1986 (Falloon)
Lea-Francis Story, The (Price)
Lexus Story, The (Long)
little book of smart, the (Jackson)
Lola – The Illustrated History (1957-1977) (Starkey)
Lola – All The Sports Racing & Single-seater Racing Cars 1978-1997 (Starkey)
Lola T70 – The Racing History & Individual Chassis Record 4th Edition (Starkey)
Lotus 49 (Oliver)
Marketingmobiles, The Wonderful Wacky World Of (Hale)
Mazda MX-5/Miata 1.6 Enthusiast's Workshop Manual (Grainger & Shoemark)
Mazda MX-5/Miata 1.8 Enthusiast's Workshop Manual (Grainger & Shoemark)
Mazda MX-5 Miata: The Book Of The World's Favourite Sportscar (Long)
Mazda MX-5 Miata Roadster (Long)
MGA (Price Williams)
MGB & MGB GT– Expert Guide (Auto-doc Series) (Williams)
MGB Electrical Systems (Astley)
Micro Caravans (Jenkinson)
Micro Trucks (Mort)
Microcars At Large! (Quellin)
Mini Cooper – The Real Thing! (Tipler)
Mitsubishi Lancer Evo, The Road Car & WRC Story (Long)
Montlhéry, The Story Of The Paris Autodrome (Boddy)
Morgan Maverick (Lawrence)
Morris Minor, 60 Years On The Road (Newell)
Moto Guzzi Sport & Le Mans Bible (Falloon)
Motor Movies – The Posters! (Veysey)
Motor Racing – Reflections Of A Lost Era (Carter)
Motorcycle Apprentice (Cakebread)
Motorcycle Road & Racing Chassis Designs (Noakes)
Motorhomes, The Illustrated History (Jenkinson)
Motorsport In colour, 1950s (Wainwright)
Nissan 300ZX & 350Z – The Z-Car Story (Long)
Off-Road Giants! – Heroes of 1960s Motorcycle Sport (Westlake)
Pass The Theory And Practical Driving Tests (Gibson & Hoole)
Peking To Paris 2007 (Young)
Plastic Toy Cars Of The 1950s & 1960s (Ralston)
Pontiac Firebird (Cranswick)
Porsche Boxster (Long)
Porsche 964, 993 & 996 Data Plate Code Breaker (Streather)
Porsche 356 (2nd Edition) (Long)
Porsche 911 Carrera – The Last Of The Evolution (Corlett)
Porsche 911R, RS & RSR, 4th Edition (Starkey)
Porsche 911 – The Definitive History 1963-1971 (Long)
Porsche 911 – The Definitive History 1971-1977 (Long)
Porsche 911 – The Definitive History 1977-1987 (Long)
Porsche 911 – The Definitive History 1987-1997 (Long)
Porsche 911 – The Definitive History 1997-2004 (Long)
Porsche 911SC 'Super Carrera' – The Essential Companion (Streather)
Porsche 914 & 914-6: The Definitive History Of The Road & Competition Cars (Long)
Porsche 924 (Long)
Porsche 944 (Long)
Porsche 993 'King Of Porsche' – The Essential Companion (Streather)
Porsche 996 'Supreme Porsche' – The Essential Companion (Streather)
Porsche Racing Cars – 1953 To 1975 (Long)
Porsche Racing Cars – 1976 On (Long)
Porsche – The Rally Story (Meredith)
Porsche: Three Generations Of Genius (Meredith)
RAC Rally Action! (Gardiner)
Rallye Sport Fords: The Inside Story (Moreton)
Redman, Jim – 6 Times World Motorcycle Champion: The Autobiography (Redman)
Rolls-Royce Silver Shadow/Bentley T Series Corniche & Camargue Revised & Enlarged Edition (Bobbitt)
Rolls-Royce Silver Spirit, Silver Spur & Bentley Mulsanne 2nd Edition (Bobbitt)
RX-7 – Mazda's Rotary Engine Sportscar (Updated & Revised New Edition) (Long)
Scooters & Microcars, The A-Z Of Popular (Dan)
Scooter Lifestyle (Grainger)
Singer Story: Cars, Commercial Vehicles, Bicycles & Motorcycle (Atkinson)
SM – Citroën's Maserati-engined Supercar (Long & Claverol)
Subaru Impreza: The Road Car And WRC Story (Long)
Supercar, How To Build your own (Thompson)
Taxi! The Story Of The 'London' Taxicab (Bobbitt)
Tinplate Toy Cars Of The 1950s & 1960s (Ralston)
Toyota Celica & Supra, The Book Of Toyota's Sports Coupés (Long)
Toyota MR2 Coupés & Spyders (Long)
Triumph Motorcycles & The Meriden Factory (Hancox)
Triumph Speed Twin & Thunderbird Bible (Woolridge)
Triumph Tiger Cub Bible (Estall)
Triumph Trophy Bible (Woolridge)
Triumph TR6 (Kimberley)
Unraced (Collins)
Velocette Motorcycles – MSS To Thruxton Updated & Revised (Burris)
Virgil Exner – Visioneer: The Official Biography Of Virgil M Exner Designer Extraordinaire (Grist)
Volkswagen Bus Book, The (Bobbitt)
Volkswagen Bus Or Van To Camper, How To Convert (Porter)
Volkswagens Of The World (Glen)
VW Beetle Cabriolet (Bobbitt)
VW Beetle – The Car Of The 20th Century (Copping)
VW Bus – 40 Years Of Splitties, Bays & Wedges (Copping)
VW Bus Book, The (Bobbitt)
VW Golf: Five Generations Of Fun (Copping & Cservenka)
VW – The Air-cooled Era (Copping)
VW T5 Camper Conversion Manual (Porter)
VW Campers (Copping)
Works Minis, The Last (Purves & Brenchley)
Works Rally Mechanic (Moylan)

First published in October 2008 by Veloce Publishing Limited, 33 Trinity Street, Dorchester DT1 1TT, England. Fax 01305 268864/e-mail info@veloce.co.uk/web www.veloce.co.uk or www.velocebooks.com.
ISBN: 978-1-84584-181-2/UPC: 6-36847-04181-6
© Graham Robson and Veloce Publishing 2008. All rights reserved. With the exception of quoting brief passages for the purpose of review, no part of this publication may be recorded, reproduced or transmitted by any means, including photocopying, without the written permission of Veloce Publishing Ltd. Throughout this book logos, model names and designations, etc, have been used for the purposes of identification, illustration and decoration. Such names are the property of the trademark holder as this is not an official publication.
Readers with ideas for automotive books, or books on other transport or related hobby subjects, are invited to write to the editorial director of Veloce Publishing at the above address.
British Library Cataloguing in Publication Data - A catalogue record for this book is available from the British Library. Typesetting, design and page make-up all by Veloce Publishing Ltd on Apple Mac.
Printed in India by Replika Press.

RS Cosworth
& World Rally Car

Ford Escort

RALLY GIANTS™

Graham Robson

Contents

Foreword 5
Introduction & acknowledgements 7
The car and the team 9
    Inspiration 9
    The Escort RS Cosworth's importance in rallying 11
    Four-wheel drive – the breakthrough 13
    Facing up to rival cars 16
    Homologation – meeting the rules 18
    Engineering features 21
    World Rally Car 27
    Escort World Rally cars – how many made, when and where? 32
    Rally car development and improvements 34
    Building and running the works cars 38
    Personalities and star drivers 42
Competition story 50
    1993 54
    1994 74
    1995 87
    1996 – The only way was up 92
    1997 100
    1998 114
    1999 119
World Rally success 120
Works rally cars (and when first used) 122
Index 124

Foreword

What is a rally? Today's events, for sure, are completely different from those of a hundred or even fifty years ago. What was once a test of reliability is now a test of speed and strength. What was once a long-distance trial, is now a series of short-distance races.

In the beginning, rallying was all about using standard cars in long-distance road events, but by the 1950s the events were toughening up. Routes became rougher, target speeds were raised, point-to-point speed tests on special stages were introduced, and high-performance machines were needed to ensure victory.

Starting in the late 1950s, too, teams began developing extra-special versions of standard cars, these were built in small numbers, and were meant only to go rallying or motor racing. These 'homologation specials' now dominate the sport. The first of these, unquestionably, was the Austin-Healey 3000, and the latest is any one of the ten-off World Rally Cars which we see on our TV screens or on the special stages of the world.

Although rally regulations changed persistently over the years, the two most important events were four-wheel drive being authorised from 1980, and the 'World Rally Car' formula (which required only 20 identical cars to be produced to gain homologation) being adopted in 1997.

At all times, however, successful rally cars have needed to blend high-performance with strength and reliability. Unlike Grand Prix cars, they have needed to be built so that major repairs could be carried out at the side of the road, in the dark, sometimes in freezing cold, and sometimes in blazing temperatures.

Over the years, some cars became dominant, only to be eclipsed when new and more advanced rivals appeared. New cars appeared almost every year, but dramatically better machines appeared less often. From time to time, rally enthusiasts would be astonished by a new model, and it was on occasions like that when a new rallying landmark was set.

So, which were the most important new cars to appear in the last half century? What is it that made them special at the time? In some cases it was perfectly obvious – Lancia's Stratos was the first-ever purpose-built rally car, the Audi Quattro was the first rally-winning four-wheel drive car, and the Toyota Celica GT4 was the first rally-winning four-wheel drive Group A car to come from Japan.

But what about Ford's original Escort? Or the Fiat 131 Abarth? Or the Lancia Delta Integrale? Or, of course, the Subaru Impreza? All of them had something unique to offer at the time, in comparison with their competitors. They offered something different, and raised rallying's standards even further; they were true Rally Giants.

To a rallying petrol-head like me, it would have been easy to choose twenty, thirty or even more rally cars that have made a difference to the sport. However, I have had to be brutal, and cull my list to the very minimum. Listed here, in chronological order, are the 'Giant' cars I have picked out, to tell the ongoing story of world-class rallying in the last fifty years:

Car	Period used as a works car
Austin-Healey 100-Six and 3000	1959-1965
Saab 96 and V4	1960-1976
Mini Cooper/Cooper S	1962-1970
Ford Escort Mk I	1968-1975
Lancia Stratos	1974-1981
Ford Escort Mk II	1975-1981
Fiat 131 Abarth	1976-1981
Audi Quattro and S1	1981-1986
Peugeot 205T16	1984-1986
Lancia Delta 4x4/Integrale	1987-1993
Toyota Celica GT4	1988-1995
Ford Escort RS Cosworth/WRC	1993-1998
Mitsubishi Lancer Evo	1995-2001
Subaru Impreza Turbo/WRC	1993-2006
Peugeot 206WRC	1999-2003
Ford Focus WRC	1999-2005

There is so much to know, to tell, and to enjoy about each of these cars that I plan to devote a compact book to each one. And to make sure that one can be compared with another, I intend to keep the same format for each volume.

Graham Robson

Introduction & acknowledgements

By any standards, the evolution of the Escort RS Cosworth was a copybook way of developing a great rally car. Not only did the inspiration for this Rally Giant come from the right source – Ford Motorsport – but so did the basic engineering, the rationale, and the development improvements laid down for the programme.

With the exception of elements of the styling, and various showroom derivatives, here was a car which Motorsport had conceived, requested, engineered, and refined, to tackle one job and one job only – to become a world-beating rally car. It was not a car, in other words, which was to be diluted by marketing or sales gimmicks – although as more than 7000 such cars would eventually be sold, those departments were not complaining. Like the RS200 (which had been killed off by regulation, rather than by its own failings) it was the sort of car which Ford Motorsport knew that it wanted, rather than one imposed on it by manufacturing realities.

Although it took a great deal of time to bring this car from Good Idea to homologated Group A machine – no less than four and a half years, much of which was caused by production practicalities – the wait was worth it. Not only did the original Escort RS Cosworth win five World Championship rallies in 1993, its first full season, but it soon became the car to use at European Championship and national level. The proof of this pudding was that between 1993 and 1996 (when the privately-prepared Escort RS Cosworth was at the height of its powers) there were no fewer than 95 European outright victories.

Even the stop-gap Escort World Rally Car, which fitted the bill in 1997 and 1998, was a winner at first, and fully competitive in the second year. Only once, in 1995, when the organisation, and the general will to win seemed to evaporate for a time, did the car look like becoming prematurely obsolete. Even then, a great deal of hard work, and the charisma of one Carlos Sainz, soon turned that situation round.

This all came about in the late 1980s because Ford Motorsport's most powerful characters – notably Stuart Turner, Peter Ashcroft, Mike Moreton and John Wheeler – were thoroughly tired of playing 'catch-up' with their competitors. When the Audi Quattro moved the goalposts in the early 1980s, Ford had no four-wheel drive car to match them; when Group A was imposed on rallying in 1987, it still had no four-wheel drive contender; and when the Lancia Delta Integrale was at its height, in 1990-1991, the Sierra Cosworth 4x4 was both too heavy and unreliable.

The original legend – now re-confirmed as fact – is that in 1988 Motorsport's 'Famous Four' sat down and thought ahead. They decided on the sort of four-wheel drive rally car they really needed to beat their opposition, and cold-bloodedly set out to sell that idea to their management. Not content, incidentally, to fight just a paper battle, they built up an attractive Escort/Sierra-based 'mule,' in which the Sierra Cosworth 4x4 platform and running gear figured strongly, and proved their point to anyone who could be persuaded to drive it.

As with the Group B RS200 project which had come five years earlier, Motorsport steadily bulldozed the project through Ford's labyrinthine layers of management, all the time pushing ahead with detail design, rally car development, and, of course, turning the closely-related Sierra Cosworth 4x4 into a formidable rally car too.

Several factors combined to make this a project which Ford just knew would succeed. First of all, motorsport suppliers like Cosworth and Mountune (engines) and FF Developments (transmissions) enjoyed real continuity at last – much of what went into the Escort RS Cosworth had already been proved in the Sierra Cosworth 4x4 – and the choice of Karmann of Germany, to produce the production cars, proved to be quite inspired. The fact that 'only' 2500 cars had to be built to meet FIA Group A requirements, made it so much easier for Karmann to do such a job – for neither Saarlouis (Ford's specialised Escort plant) or Genk (Sierras) could have accepted such a limited production challenge.

As far as Boreham was concerned, the Escort RS Cosworth came along at exactly the right time. Not only had the department finally honed the Sierra Cosworth 4x4 into a great, if not Championship-winning rally car, but its opposition, notably the Lancia Delta Integrale, had run out of development potential. For a time, the combination of Escort RS Cosworth capability, Francois Delecour's unearthly driving skills, and Boreham's gritty resolve, made this the operation which set all the standards.

Unhappily, in 1994 and 1995, a series of hard-faced management decisions at the top level in Ford, and the ruthless cutting of some funds, meant that the works team struggled for a time. It was only luck, and, some say, sheer bloody-minded determination, which allowed the Escort RS Cosworth to become competitive again in 1996.

The development of the Escort World Rally Car, which completes this story, is another lesson in how to meet stifling regulations head on, and how to make much of very little in no time at all. Other than Ford, how many teams could have produced such an improved car in such a short time, using little more than rallying petty cash to achieve the impossible?

During a compact, six year career, not only was the Escort RS Cosworth/World Rally Car family a great success, but it maintained the Escort's motor sporting heritage until that family was more than thirty years old. In every way, therefore, it was a true Rally Giant.

Acknowledgements

As ever, I want to thank many people for helping me to assemble all the facts which are in this book. Going back over the years – I made my first visit to Boreham in 1965, and was in and out of those premises for the next forty years – it was personalities like Stuart Turner, Peter Ashcroft and Colin Dobinson who helped me to know so much about these Escorts, along with colleagues who are all mentioned in the text, not least Mike Moreton, John Taylor, John Wheeler and Malcolm Wilson.

Dave Hill (and his predecessors, Sheila Knapman and Fran Chamberlain) runs the Ford Photographic Archive with such efficiency that it has always been a pleasure to visit, researching the images which are in these pages. Without them, there would be much white space – and a lack of excuses from me!

Recently, and in the search for obscure cars, events and occasions, I have come to rely more and more on Martin Holmes, whose *World Rallying* annuals are a constant source of inspiration, and whose personal pictorial archive is a gold mine too. Like others in my profession, I have come to rely on Martin as an absolute authority, to be cherished at all times.

I am delighted to confirm that everything they did, and continue to do, makes writing books about Ford Motorsport as enjoyable as ever.

Graham Robson

The car and the team

Inspiration

By the late 1980s, Ford Motorsport knew that it needed a new rally car. Having spent a fortune developing the Group B RS200, it had seen this car sunk by an abrupt change of regulations. When the FIA imposed Group A rallying instead of Group B, Ford found itself without a competitive four-wheel drive car, and, struggling to match its rivals, it was forced to use the ultra powerful, rear-drive Sierra RS Cosworths.

With cars like the Lancia Delta Integrale and, shortly, the Toyota Celica GT4 making the running, Ford's only immediate answer was to await the arrival of its own four-wheel drive, the Sierra Cosworth 4x4. Unhappily, that car could not go on sale until 1990, and although a competent machine, it was too large and heavy to be truly competitive.

It was therefore imperative to produce a new model. If Ford Motorsport wanted to start winning World rallies again – and it certainly did, having only won once since the great days of the Mk II Escort came to an end – it needed a new state-of-the-art car with which to do it. The inspiration came, in 1988, from Ford Motorsport's Boreham HQ, where, as the then Motorsport Director, Stuart Turner, later told me: "Peter Ashcroft, Mike Moreton, John Wheeler, John Griffiths, Bill Meade and I used to get together regularly, for meetings about our future."

"I think we were all agreed that whatever our next rally car would be, it had to be based on the platform and the basic layout of an existing mainstream model ... This time,

Director of Motor Sport, Stuart Turner (right) and Carlos Sainz were great friends for many years.

Walter Hayes – visionary, PR genius, and all-important personality at Ford for thirty years, always made sure that the Escort RS Cosworth received its share of resources.

we wanted to plan well ahead for the 1990s, and we wanted to start at once. I think it was one of my comments, thrown into the conversation, that encouraged a breakthrough: 'Why don't we see if we can take the platform and running gear from a Sierra Cosworth 4x4, shorten it, then see if an Escort body will fit on it?'"

Romantic? Yes. Fictional? No, this really was how the Escort RS Cosworth was conceived. The elements of a new front-wheel drive hatchback shell were somehow made to fit the shortened platform of an entirely different, four-wheel drive saloon. Although this all happened in 1988, the production car would not go on sale for another four years.

Way back in the 1970s, Ford's original Escorts had won rallies all round the world. This was Hannu Mikkola on the way to victory in the 1972 Safari.

Incidentally, for many years Ford insiders knew this project as the 'ACE.' This was because the new-generation Escort, on which the hatchback bodyshell was closely based, was coded CE14, and the rally car would have to comply with Group A regulations. Clumsily, therefore, the 'Group A version of CE14' was shortened to the 'ACE' code, and it stuck.

This Boreham meeting happened, please note, two whole years before the bigger and heavier Sierra Cosworth 4x4 even went on sale, but since the personalities involved all knew how long it was going to take to convert a dream into reality, they knew they had to start soon. They also realised that the existing Lancia Delta Integrale had a real stranglehold on the sport, and would need to be toppled. One day ...

The Escort RS Cosworth's importance in rallying

In the 1990s, as far as Ford was concerned, a new car like the Escort RS Cosworth was vital to the company's rallying future. Whereas the team had set most standards in the 1970s, when the Escort Mk I and Mk II family had been so successful, in the 1980s the company had consistently lagged behind its rivals. This was not because of a lack of interest, nor for getting its timing wrong so often, nor for being what some of its critics often quipped as 'one day

Ford's Mk II Escort has probably won more rallies than any other car in the world – and when the Escort RS Cosworth was conceived, Ford intended to make a new car which would match it.

late, one dollar short.' There is no doubt that the company had been unfortunate, the exciting RS200 had been one of the last Group B cars to be homologated. There had been no immediately competitive four-wheel drive Ford available when Group A was imposed on the sport in 1987, and the Sierra Cosworth 4x4, though finally sturdy and versatile, had never been outstanding in a sport where increasingly specialised cars were needed to ensure success.

Fortunately, by 1988, there were individuals at Ford Motorsport – notably Stuart Turner, Peter Ashcroft, and John Wheeler – who were no longer willing to keep on playing catch-up. To close the gap, they began to plan years, rather than months, into the future, and began to design a long-term successor to the Sierra Cosworth 4x4, which had still, incidentally, not even been revealed to the public! This is where the original ACE project was born.

The idea of the new ACE germinated in 1988, the first car was finished in 1990, sales of road cars began in

1992, and Group A homologation was achieved on 1 January 1993. Each and every one of those dates was critical to the programme, but there were other, equally vital, reasons for this being one of the most important Ford rally cars of all time.

It was generally agreed that Ford's standing, if not its influence, in top class rallying had been eroded in the 1980s. No sooner had the family of rear-wheel drive Escorts been retired, than the brand became less and less attractive to a mass of rally teams and drivers all around the world. As fewer and fewer Fords figured in entry lists (and particularly in the results tables) of rallies, so the company's sporting reputation began to crumble.

Accordingly, for those with loyalties to Ford in rallying, it was vitally important that a new compact car, like the Escort RS Cosworth, should succeed. Not only had Ford been the only British marque consistently present at the very top level of the sport, but it was also one of the few companies which consistently made all its technology, and its developments, available to private owners. In the 1970s, there were more competitive Fords entering national and international rallies to European level than any other car, and Ford thought it important to aim for that again in the 1990s.

Introduced in 1984, briefly enjoying Group B rallying in 1986 then going on to become an extremely successful rallycross car, the four-wheel drive RS200 was conceived entirely at Boreham. The Escort RS Cosworth was to be a worthy successor to this machine.

Four-wheel drive – the breakthrough

Until the 1980s, the use of four-wheel drive was strictly forbidden in international rallies. Not that this distressed many people. At this time, the only four-wheel drive cars to be on the market were under-powered Subarus from Japan. Then came the revolution, when Audi of Germany not only took an interest in rallying, but also began to develop a high-performance, turbocharged coupé, the Quattro, which only needed four-wheel drive to complete an exciting package.

When Audi then started using the brutish new Quattro, in 1981, with Hannu Mikkola and Michele Mouton as its works drivers, the sport of rallying changed forever. For the very first time, rallying saw that four-wheel drive had been harnessed to a powerful chassis, a combination which soon meant that special stage times were slashed. Although there would be a short period in which conventional two-wheel drive cars could still win on sealed surfaces, any manufacturer with long-term rallying aims had to start looking at four-wheel drive.

The Sierra Cosworth 4x4 of 1990-1992 was Ford's first four-wheel drive Group A car. Slightly too heavy and too bulky, it would bequeath its basic platform and running gear to the new ACE-Escort RS Cosworth project.

Over at Boreham, Ford evolved the ultra-special RS200, but soon found it rendered obsolete. Following the horrifying accidents which befell spectators in Portugal, and Henri Toivonen in Corsica, the FIA cancelled the Group B category, and imposed Group A instead, where four-seaters had to be built in quantities of at least 5000 to be homologated.

At such short notice, Ford was instantly disadvantaged. At that time, its only four-wheel drive car was the Sierra XR 4x4, which was not only too big and too heavy, but also had an asthmatic V6 engine. Lobbying at director level soon saw a four-wheel drive Sierra Cosworth 4x4 under development, but that could not be put to use before mid-1990, and even then, the problem was that it was still too big and too heavy to be able to beat the more nimble Lancia Delta Integrale, which was setting all the standards at that time.

As we all now know, the Sierra Cosworth 4x4 four-wheel drive installation, suitably updated and refined, was

By 1992, the works Sierra Cosworth 4x4 was reliable, with high performance and excellent handling, but was just a touch too bulky to be a consistent winner. The Escort RS Cosworth which replaced it would be a better car in all respects.

adopted for the new Escort RS Cosworth, and this was a real breakthrough for Ford. Not only was it to be fitted to a car which had at least as much, and maybe more, power than the Sierra Cosworth 4x4 had ever had, but it was also one of the most sophisticated and versatile 4x4 installations currently available in private cars.

Because of the modular way in which Ford's mainstream engineers had laid out the Sierra/Escort type of four-wheel drive, it meant that the Motorsport engineers at Boreham could develop new and more specialised versions of various bits and pieces for their own use. Although the original installation – front engine/central main gearbox/separate centre differential/separate front and rear differentials – was always retained, before long there were different Motorsport

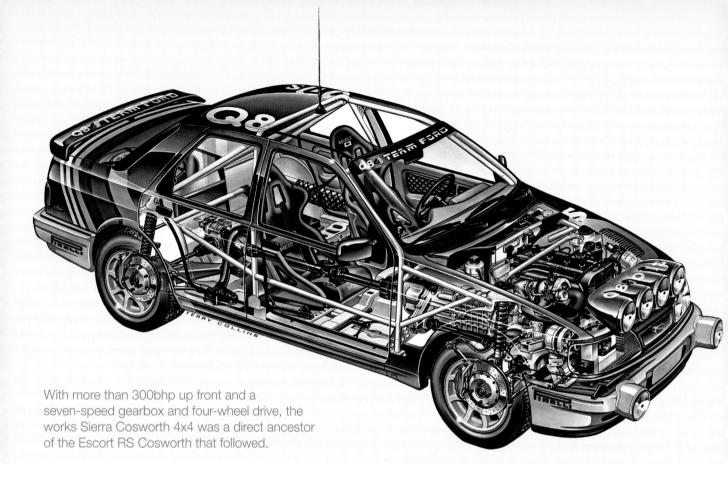

With more than 300bhp up front and a seven-speed gearbox and four-wheel drive, the works Sierra Cosworth 4x4 was a direct ancestor of the Escort RS Cosworth that followed.

versions of all the separate sub-assemblies. That was the real breakthrough which made the development, improvement and evolution of the Escort RS Cosworth and, later, the Escort World Rally Car (WRC), so satisfying.

Facing up to rival cars

When the Escort RS Cosworth was conceived, the most serious competition came from Lancia, with the Delta Integrale, and Toyota, with Celica GT4 types. Even by 1992, when series production began, these were still the standard bearers which Ford would have to beat when it started rallying the following year. Although by then, Group A regulations had been modified – in 1992 it had only been necessary to build more than 2500 cars instead of 5000. (Ford, along with Karmann, was well-equipped to meet either target.) Cars being designed to take on Lancia, Toyota and now Ford, would all have to be seriously developed machines.

Ford was confident that the Escort RS Cosworth could eventually beat Lancia and Toyota, but, in the next few years, it would also have to face up to the Mitsubishi Lancer Evolution and the Subaru Impreza Turbo, both of which were backed by proud Japanese masters equipped with more funds than Ford could even dream about.

Then, in 1997, when the rules of rallying changed yet again, Ford had to gain a rule-bending dispensation from the authorities, and develop the Escort World Rally Car in a great hurry to face up to the latest Japanese cars. Toyota was also starting again, with the compact Corolla World Rally Car, and other new models would surely follow.

This is how the main competition stacked up while the Escort RS Cosworth and World Rally Car models were current:

Lancia Delta Integrale – front transverse-engine/four-wheel drive. The established, long term favourite, with a success record stretching back to 1987 when it was the less powerful, Delta HF 4x4. A multiple World Champion in the Makes series, and for the lucky drivers hired to use it. Perhaps a touch small and lacking in wheel movement, but powerful, the lightest, most nimble, and best financed of all. With works support withdrawn at the end of 1992, it was now past its peak. After recording eight World victories in 1992, there would be only two second places in 1993, and none after that.

Mitsubishi Lancer Evolution – front transverse-engine/four-wheel drive. The earlier Galant had been too heavy and too large to be a consistent winner, nor was it backed by a full-blooded, fully-financed programme. From 1993, the new and smaller Lancer Evolution was more competitive, especially on loose-surface, rough-road events when driven by top-grade drivers, although these were delayed due to parsimonious management policies. Mitsubishi, like Subaru, was a team for the future, still to build a reputation. Galants had won six World events by 1992, most of them on endurance events. From 1995, though, its Unique Selling Proposition (USP) was that Tommi Makinen would join as its lead driver. It was not until he left – actually to drive for Subaru – that the sequence ended.

Toyota Celica Turbo 4WD – front transverse-engine/four-wheel drive. As the Delta Integrale went into decline, the latest Celica, first homologated on 1 January 1992, was the standard setter. Powerful, well-engineered (by TTE in Cologne) and seemingly with a limitless budget, the Celica team had won five World rounds in 1992, and would win seven more in 1993. The drivers – Juha Kankkunen and Didier Auriol – were first rate, and the team, ruthless. Widespread rumours of rule bending were rarely pursued by scrutineers, though the Celica's astonishing straight-line performance could never be explained by normal analysis.

Retribution would follow in the mid-1990s when the team cheated, was caught red-handed, and was banned from World events for the whole of 1996. No-one in rallying suggested that this punishment was too harsh.

Toyota Corolla WRC – front transverse-engine/four-wheel drive. This was designed and developed by TTE in Germany while Toyota was banned from rallying. Smaller, lighter and more nimble than the old Celica, the Corolla mated a new, family-car hatchback shell with the old-type Celica engine/transmission. First seen in mid-1997 and homologated in August 1997, it was a direct competitor to the Escort World Rally Car, and although initially it had little impact, it was successful in 1998 with Sainz and Auriol as team drivers. Second in the World Makes Championship in 1998, Sainz lost the Drivers' series only at the end, when an engine blew on the RAC Rally. Toyota was very serious about this programme, using no fewer than 36 individual cars in 1998. A formidable team, generously financed, with the best drivers.

Subaru Impreza Turbo WRC – flat-four front engine/four-wheel drive. Here was truly formidable competition, maturing at the same time as the Ford. Based at Banbury in the UK, Prodrive, the vastly experienced motor sport consultancy concern, had been running the Japanese Subaru's World Rally effort since 1990. With an engine, transmission and four-wheel drive layout based on that of the bigger, older, Legacy model, the Impreza was immediately competitive. Its first World win would come in 1994, and Colin McRae would become World Champion in 1995. Along with the Mitsubishi Lancer, the Subaru and the Subaru Impreza WRC which followed were the cars which Ford had to beat.

By 1998, Ford Motorsport knew that a replacement for the Escort was now close to launch, and that, for image reasons, it would be expected to go rallying with facsimiles of that new range. Accordingly, engineering development of Escort WRCs was much reduced, the cars ending their career at the end of the year almost in the same state that they had been in at the beginning of the season. Engineering work on a new Focus WRC, which had absolutely nothing in common with the Escort WRC, began early in 1998, the new rally car then being unveiled to the public just before the end of the year.

It had been hoped that the Escort WRC would never be used again after 1998, but there was a single entry in the 1999 Swedish Rally before this story did, indeed, come to a close.

Homologation – meeting the rules

Well before the Escort RS Cosworth was finalised, Motorsport realised that it could not cut any corners to achieve Group A homologation. All around the world, so many doubtful homologation approvals had been granted in recent years (most, but not all, Group B cars, had got into the sport well before the requisite number of cars had been constructed) that the authorities began to insist on really detailed chapter-and-verse applications. Not only that, but they wanted to be convinced that the necessary number of cars had been built.

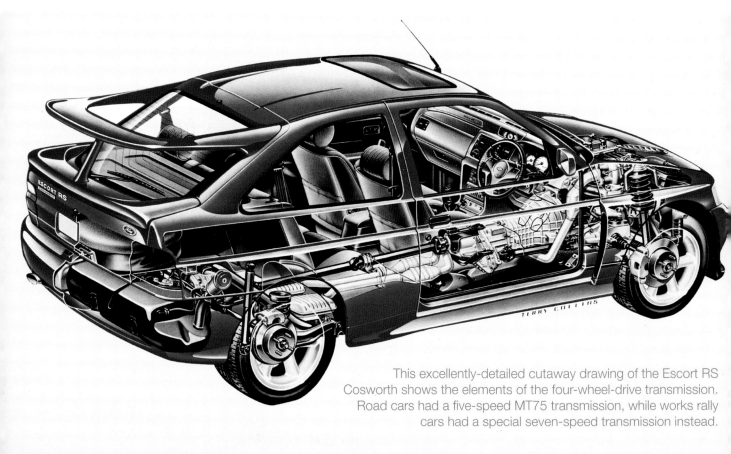

This excellently-detailed cutaway drawing of the Escort RS Cosworth shows the elements of the four-wheel-drive transmission. Road cars had a five-speed MT75 transmission, while works rally cars had a special seven-speed transmission instead.

Not what it seems! This, in fact, was the very first 'mule' for the ACE-Escort RS Cosworth project, as under the modified old-type Escort body was a shortened Sierra Cosworth 4x4 platform. Engineer, John Wheeler (right) and project manager Mike Moreton pose with the car they had conceived in 1988.

Though this was never intended for motorsport use, from mid-1994 the road car engine was thoroughly re-worked with a smaller turbocharger and Ford EECIV electronic engine management.

When the ACE project took shape in 1988/1989, 5000 cars needed to be built for Group A approval to be granted, but from 1991 the regulations were eased, and this figure had been halved, to 2500, which made a manufacturer's job considerably easier. Even so, when preparation for ACE/Escort RS Cosworth series production began, Ford had to plan for 5000 cars.

Because this car was very special in so many ways, the company soon concluded that it could not be assembled on the same tracks in Europe where other mass-production Fords were taking shape. After a lot of investigation had taken place (Boreham's Mike Moreton was much involved in this) it was decided to place a contract with the German independent concern, Karmann, for the assembly job.

This was much more logical that it sounded at first. Not only was Karmann an extremely resourceful maker of bodyshells, but it could, and did, undertake complete car assembly for other concerns. One of its very first enterprises was the Karmann-Ghia coupé, which was based on the VW Beetle, but in more modern times it had also got involved in making thousands of bodyshells for the Porsche 924 project, and in undertaking complete assembly for the VW Golf GTI Cabriolet and Ford Escort Cabriolet projects.

The most important factor, though, was that Karmann had undertaken complete assembly of the Sierra-based Merkur XR4Ti model for Ford, and, by the end of the 1980s, that activity was coming to an end. This left Karmann with a factory at Rheine, in the north-west of the country, which was yawningly empty in some respects, and a desire to start building a new car as soon as possible.

The deal with Karmann was concluded as early as 1989, when no car had yet been produced, though by then the company had already secured a deal to produce new-generation Escort Cabriolets, in their entirety, from 1991.

To meet the letter of the FIA's Group A regulations, every one of the first 2500 examples had the familiar front splitter/high rear aerofoil package. To complicate matters, however, there was only one mechanical specification but three different trim/equipment levels for sale throughout Europe. Later, there was a version without the rear aerofoil.

Well before the end of 1992, however, the rush to buy was over, for in the UK at least, the insurance industry did its very best to kill off the car's appeal by loading premiums to ludicrous and quite unjustified levels. We now know that nearly 3500 were built between February 1992 (when the first Geneva Show cars and the first series-production cars were assembled) and the end of that year. True production of the series began on 27 April 1992, and the 'on-sale' date in the UK was 22 May 1992. However, by 1994, Karmann was only building little more than 1100 cars a year.

Then, with sporting homologation finally settled, the Mark 2 Escort RS Cosworth appeared in mid-1994, slightly less extrovert, better developed, and yet more driveable, with the more flexible, small-turbo YBP engine. These are the official, year-on-year production figures:

1992 3448
1993 1143
1994 1180
1995 1306
1996 68

Then, in September 1995, Ford announced that the Escort RS Cosworth would be killed off in January 1996, officially because it was no longer capable of meeting the latest exhaust emission limits, or the new 'drive-by' noise limits. Some suggested that this was spurious reasoning, and that it was poor sales which really brought about the cancellation. In the end, as can be seen, no fewer than 7145 road cars had been built – which made it one of the mostly honestly launched, produced and rallied Group A 'super-cars' the world had ever seen.

Engineering features

To win in motorsport, more than 300bhp with four-wheel drive would be needed. As a rally car it had to be as suitable for the Safari as the RAC Rally, for Corsica as well as the Monte. It needed to be better, a lot better, than the Lancia Integrale, and capable of improvement in the years which followed. And in basic form, it had to be a practical road

The double spoiler at the rear gave a great deal of positive aerodynamic downforce at higher road speeds.

car – Ford would not approve the production of 5000 identical cars if they could not be sold, or make a profit.

John Wheeler's analysis was masterly. Rejecting as a basis the new-type Fiesta (physically too small to accommodate the in-line engine, transmission and massive wheels that would be needed), and the forthcoming Mondeo (too large, and its launch too far into Ford's future), he settled on the cabin 'package' of the next Escort.

He determined to have an aerodynamically stable car – and if this meant that it looked more extreme than the Sierra RS Cosworth, so be it. The vast rear spoiler which resulted was not purely a styling exercise, for Ford's German wind tunnels were occupied for 200 hours, and took account of drag, down force, and engine-bay air flow.

Although the Lancia and other current rally cars all used transverse-engine layouts, Wheeler rejected that idea completely. An in-line engine, with a gearbox behind it, offered more, not only in terms of accessibility, but also in ideal weight distribution. For such a new car, developed versions of the Sierra Cosworth 4x4's platform and proven running gear would be ideal. It was happenstance, not sheer good luck, that the Sierra running gear could be adapted to fit so well, and allowed development time to be telescoped.

The first prototype was built on an unofficial basis. Though the Sierra Cosworth 4x4 was still more than a year away from production, an engine and a four-wheel drive transmission from that car, along with the floor pan and suspension of a Sierra XR 4x4, were somehow procured.

In certain territories the Escort RS Cosworth was available without the roof-top spoiler – but very few such cars were sold.

With expert help from TC Prototypes, the RS Turbo-based ACE 14 'mule' was completed in a matter of weeks. John Thompson's team shortened the Sierra floor pan, inserted the RS Cosworth engine and transmission, grafted the modified old-type Escort RS Turbo body superstructure on to it – and the original ACE was born.

It then took years to turn a Good Idea into a production car. The Escort RS Cosworth rally car was officially previewed in September 1990, when Mia Bardolet won the Spanish Talavera Rally on its world debut. Production cars went on sale in May 1992, but motorsport homologation was not achieved until 1 January 1993. Once approved, the project was turned over to Special Vehicle Engineering (SVE) for completion. John Wheeler moved to Dunton as the Chief Project Engineer, and never more than a dozen or so SVE personnel worked on the project at the same time.

Inevitably, the original romantic notion – that a new-type Escort bodyshell could be enterprisingly cut-and-shut on a slightly shortened Sierra Cosworth 4x4 platform and rolling chassis – had to be modified, as many detailed structural and aerodynamic imperatives were identified. In the end, the Escort's doors, roof, hatchback and glass were still recognisable, but the front and rear wings, bonnet and other major pressings were all changed or unique: 50 per cent of the bodyshell was new, and there were 400 unique components.

As an example, Escort RS Cosworth tyres were so bulky that they could not be fitted into the existing spare wheel (from the Sierra Cosworth 4x4 floor pan), so, on road cars, the only spare provided was a puny get-you-home tyre on a steel rim that was rated only for 50mph and looked quite ridiculous when fitted.

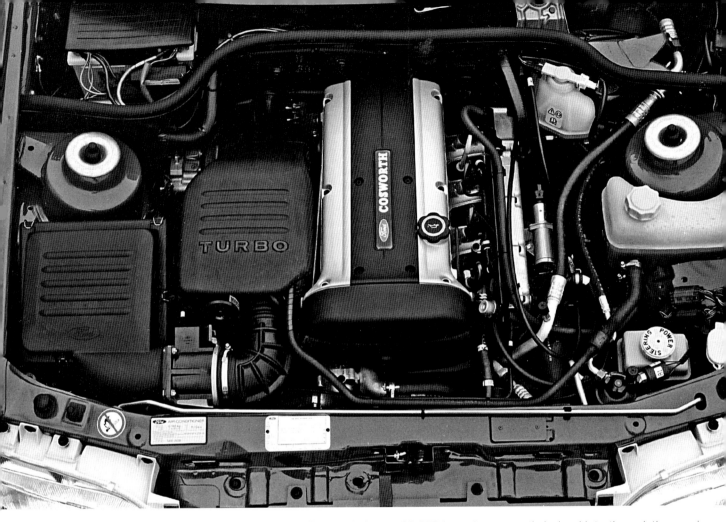

The small turbo engine fitted to the Escort RS Cosworth from mid-1994 went very neatly indeed into the existing engine bay.

SVE oversaw the development of a more powerful version of the famous YB engine. Although it might have looked similar to that of the Sierra Cosworth 4x4, it *had* advanced, notably with a hybrid TO3/T04B Garrett AiResearch turbocharger (which was really too big for road cars, but Motorsport needed a unit that big so that the engine would be useful for full-house Group A tuning ...). With 227bhp at 6250rpm – and with a limited duration overboost feature for overtaking – this Cosworth-described YBT engine was an extremely flexible power unit, and naturally it ran on unleaded fuel with an exhaust catalyst as standard.

There was much more to the styling changes than might be apparent. Almost every change made was to manage the air flow around, under, over or through the engine bay. To embrace the massive eight inch wheel rims and 225/45ZR-16in tyres, the front and rear wings had to be reshaped and

This was the instrument panel layout of the original Escort RS Cosworth road car of 1992, complete with back-lit instruments and the unique seat decoration panels.

flared, the front bumper moulding took account of optimised air flow into the engine bay, provision was made to extract hot air from the engine bay and the front brakes, and the entire package had to be made feasible for manufacture in numbers.

This was a car which developed positive aerodynamic down-force at all cruising and higher speeds – Ford claimed this as the first mass-production car to have that feature – it was 11.2in/285mm shorter than the Sierra Cosworth 4x4, with a 2.2in/56mm shorter wheelbase.

For a time, SVE had to make do with the Escort RS Turbo-based mule, while the first true prototype car was

Nothing was added to the Escort RS Cosworth for decoration only. In this side-on view, the positioning of the huge rear spoiler and the exit louvres at the rear of the front-wing pressings are obvious.

completed in February 1990. That car was by no means visually finalised, for it ran on old-style wheels, had no cooling louvres in the bonnet, and no rear spoiler. There would not be a completely representative machine until much later.

We now know that in a very compressed development schedule, just 19 prototypes were built, including a clay model used to settle the design, and more than one was subjected to the obligatory, 30mph head-on crash test. The original mule was used until mid-1990 (cold climate testing and intercooler development being among its more important tasks) and by the time the press saw the new mainstream Escort/CE14 model in August 1990, when the flame-blue liveried Escort RS Cosworth rally car made its startling debut at the Blenheim Palace launch, four Phase II Escort RS Cosworths were beginning to start work, and eight more Phase III cars were due to start testing before the end of that year.

World Rally Car

To meet the World Rally Car regulations due to take effect in 1997, Ford Motorsport needed a new model, as the Escort RS Cosworth was no longer eligible. The new rules were easy enough to meet – if a suitable base car was available. Although only 20 machines had to be built in one season (none of them for use as road cars), they had to be based on models of which at least 25,000 examples were being made in a year, such as front-wheel drive Escorts. That was the bad news. The good news, though, was that almost every other modification was allowed. Conversion from front-wheel drive to four-wheel drive was authorised, as was turbocharging, but engines had to be no more than 2-litres in capacity.

World Rally Car regulations were to be applied from 1 January 1997 – but with certain exceptions. It was the exceptions, and the opportunities, that caught Ford's attention. One big change from the past was that World Rally Cars would never be allowed to be re-shelled. If a shell was written off – specifically, if the integral roll cage was rendered useless – so was the identity of that car. Suddenly, as far as the enthusiasts were concerned, this made number plate spotting worth doing again! Spotters, therefore, will want to know that Chassis No. 004, registered P6 FMC, was the most regularly used 1997 Escort WRC, as it started no fewer than nine World Rallies.

Even before they started the job, Boreham's engineers, then led by Philip Dunabin, soon realised that a World Rally Car could eventually become very specialised. The bad news was that it would immediately make the Escort RS Cosworth obsolete.

Philip Dunabin backed John Wheeler in the original design of the Escort RS Cosworth, then took over as Motorsport's Chief Rally Engineer when John moved on to Aston Martin.

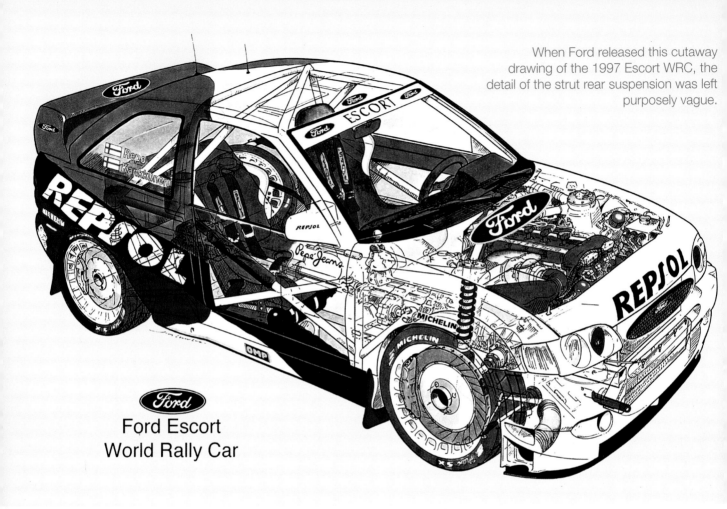

When Ford released this cutaway drawing of the 1997 Escort WRC, the detail of the strut rear suspension was left purposely vague.

Ford Escort World Rally Car

For Ford, in 1996, the biggest problem was that according to the regulations, the following year the new WRCs had to be based on a current 25,000/year model. The old Escort RS Cosworth had never approached these levels – and it was going out of production in 1996. There was no other obvious Ford model which immediately qualified. The existing Fiesta was too small, the Escort replacement – to be badged 'Focus' – would not be launched until 1998, and a competition car could not be readied before 1999.

It was time for the FIA's bluff to be called – and, as one of the few manufacturers currently dedicated to a full World Rally programme, Ford decided to do just that. Ford Motorsport therefore told the FIA that building an all-new WRC was out of the question – but offered a short-term solution which would need the acquiescence of rival manufacturers to make it feasible. If the FIA would let Ford's first-generation WRC be based on the old Escort RS Cosworth, then a team of works Fords could be on the starting line for the Monte Carlo Rally, in January 1997. And if not, then Ford would have to withdraw …

Unsurprisingly, the FIA soon agreed to this, after which the rush to get a competitive Escort WRC designed, developed and ready began at high speed. Boreham did all the design and original development work, though, as we now know, Malcolm Wilson's M-Sport would eventually get the contract to run the team cars from the start of 1997.

Concept engineering work began in June 1996, and the first prototype ran on 13 October. The launch to the media followed on 3 November, and the homologation inspection of the 20 kits of parts was completed on 19 December! There was such a rush that Motorsport at Boreham only built two prototype test cars – N704 FAR (tarmac-spec) and M513 WJN (gravel-spec) – both of them originally works Group A Escort RS Cosworth rally cars. That was a super-rapid programme by any standards, for the Escort WRC went from Good Idea to homologation in six months!

When the car was unveiled, Engineering Manager Philip Dunabin commented: "Our request to the FIA to base the World Rally Car on the existing Escort 4x4 has allowed us to build a long-term plan on the existing and future Escort models ... the decision to base the World Rally Car on the current Escort was never really questioned."

In evolving the Escort WRC from the Escort RS Cosworth, Motorsport made the following basic improvements, some of which had already been tested in the past, though naturally not on World Rallies where homologation would have been required:

Engine: New, smaller, to suit a 34mm restrictor, IHI turbocharger, different exhaust manifold and fuel injection changes with eight instead of four active injectors. This delivered 310bhp at 5500rpm, with a very solid torque curve.

Cooling: Better airflow through larger front apertures, with relocated and more efficient intercooler and water radiators.

There was continuous development in the engine bay of the RS Cosworth/WRC cars – this being one of the WRC models used in 1997.

Developed in a great hurry in the summer of 1996, the Escort World Rally Car was finally previewed in Spain in November of that year. At the front of the car there was a different grille, a huge front spoiler, and (just visible) a larger air/air engine intercooler.

Rear suspension: Lighter and stronger tubular sub-frame, a new MacPherson strut/links system with a geometry rather like that of the Mondeo.

Aerodynamics: New front bumper profile to suit the revised radiator/intercooler relocation. Smaller, reshaped rear aerofoil to generate more downforce with less drag.

Layout: Idealised, with an 80-litre fuel tank, spare wheel, and 40-litre water reservoir (water to be available for cooling sprays of intercooler and brakes) positioned in the rear compartment.

The aerodynamic tests were carried out in the Ford tunnel at Merkenich in Germany, eventually resulting in improved down-force. In any case, the new rear aerofoil was needed to satisfy WRC regulations, as the existing RS Cosworth style was too large.

The water radiator was 33 per cent larger than before, the turbo intercooler was 50 per cent larger, and it was also cooled by water spray from the reservoir: this intercooler was also re-positioned, now being ahead, rather than on top of, the radiator.

For the very first time, the works cars would be run by an outside agency. Therefore, at the end of 1996, Malcolm Wilson's Cumbrian-based M-Sport team took over. Motorsport at Boreham takes the credit for the design and development of the Escort WRC, and for the hurried manufacture of the first 20 sets of components, but it was Malcolm Wilson's team which would always run the cars.

It was at this point that the Operations Manager, John Taylor, was made redundant – most unfairly according to the stories which circulated – after more than 20 years involvement with Boreham. Now the team would be run by Malcolm Wilson, with Marc Amblard as Senior Engineer.

Carlos Sainz led the drivers, and was still a real inspiration. Originally, in the early months of 1997, he was joined by the German, Armin Schwarz. Unhappily, Schwarz's sponsorship funds never arrived, so he was later dropped, and replaced by four-times World Rally Champion, Juha Kankkunen. Although Juha brought no funds with him, he was a faster driver, more experienced in loose-surface motorsport, and would become an ideal partner for Carlos Sainz.

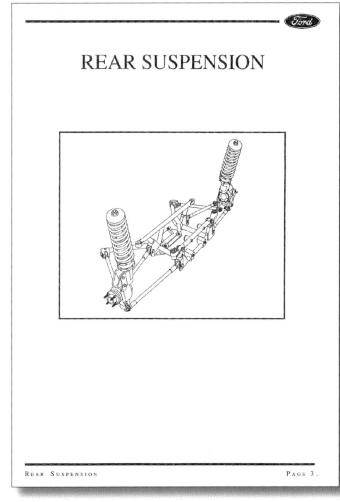

The major change which converted the Escort RS Cosworth into the Escort WRC was a new type of strut-independent rear suspension, which improved the grip, handling and balance of the car whilst using the same bodyshell pick-up points.

Compared with the earlier Escort RS Cosworth, the Escort WRC had a different rear spoiler arrangement needed to meet new WRC dimensional regulations, and a different rear bumper moulding.

In the early weeks and months of 1997, there were many cynics who suggested that M-Sport would simply not get the job done. But it did! By the middle of the year, the construction of new cars in Cumbria was almost a batch production business. By the end of the season, no fewer than 29 Escort WRCs existed – ten of them actually being converted from Group A Escort RS Cosworths of 1995 or 1996 vintage.

Somehow, M-Sport got two new cars (P6 FMC and P7 FMC) to the start line in Monte Carlo, Sweden and Portugal. The rest of the season was always a race against time, but by using only nine cars during the year, the team made it.

There were technical advances in 1997, including differential changes, with an Xtrac sequential gear change, and Hi-Tech dampers replacing Dynamic types, from mid-season.

No fewer than 19 Escort WRCs appeared in the 1998 World Championship season, of which only four had featured the previous year. Only nine different cars were official M-Sport entries. When a number of private conversions, which appeared in lesser events throughout the season, are added in to the total, this brings the final number of Escort WRCs built in two years to well over 40. Not bad for a stop gap design ...

Immediately before the Rally of Great Britain, in November 1998, Ford chose to celebrate the imminent retirement of the works Escorts by hosting a glittering celebratory dinner for every important Ford personality who could be persuaded to attend. In an occasion which has never been matched by any other team, Ford attracted no fewer than six World Rally Champions to attend – Bjorn Waldegård, Ari Vatanen, Hannu Mikkola, Juha Kankkunen, Carlos Sainz and Tommi Makinen. Timo Makinen (a triple Escort rally winner), Ove Andersson, David Richards, Malcolm Wilson and Andrew Cowan were also there. Walter Hayes, Stuart Turner, Martin Whitaker, Bill Barnett and many others joined them – and there was regret that Peter Ashcroft (who, by this time, had retired to live in the United States) could not be there. It was an honour for us all to be present, yet there was even more special pleasure in receiving a framed picture of the group of personalities, all of whom had signed their names around the margins. An unforgettable evening.

Escort World Rally Cars
How many made, when and where?

Note : Cars originally rallied in earlier years as Escort RS Cosworths are identified as such.

Identity	Built by/in	Registration (UK)	Notes/drivers/initial usage
1996:			
Boreham built two prototypes:			
001	Boreham	N704 FAR	Tarmac test car. Ex-1996 Escort RS Cosworth (Sainz 2nd/Argentina, 3rd/Australia)
002	Boreham	M513 WJN	Gravel test car. Ex-1995/1996 Escort RS Cosworth (5th and 6th placings)
1997:			
003	M-Sport	P7 FMC	Works Repsol car
004	M-Sport	P6 FMC	Works Repsol car
005	M-Sport	P11 FMC	Works Repsol car
006	M-Sport	M10MWM	[Ex-Escort RS Cosworth] Works Repsol car
007	M-Sport	M40 FMC (later P6 FMC)	[Ex Escort RS Cosworth] Works Repsol car
008	[Not completed]	-	-
009	Mike Little Preparations	P194 FAO	[Ex Escort RS Cosworth]
010	RED	P10 RED	[Ex-Escort RS Cosworth]
011	Boreham	-	-
012	M-Sport	P8 FMC	Works Repsol car
012A	M-Sport	P9 FMC	Works Repsol car
014	M-Sport	-	[Ex Escort RS Cosworth]
015	Boreham	R475 KVX	-
016	MLP	M743 YWC	[Ex Escort RS Cosworth]

Identity	Built by/in	Registration (UK)	Notes/drivers/initial usage
017	Boreham	R963 DHK	-
018	RED	-	-
019	Boreham	-	-
020	Finland	-	-
021	Boreham	-	-
022	Boreham	-	-
023	Greece	[Ex-Escort RS Cosworth]	-
024	M-Sport	-	-
025	MLP	Ex- Escort RS Cosworth]	-
026	Boreham	R771 GTW	-
027	Boreham	-	-
030	M-Sport	R2 FMC	Works Repsol car
031	M-Sport	R1 FMC	Works Repsol car
1998			
028	M-Sport	R3 FMC	1998 Works Valvoline car
029	M-Sport	R4 FMC	1998 Works Valvoline car
032	Jolly Club	R9 FMC	-
033	Jolly Club	R10 FMC	-
034	Boreham/MLP	-	-
035	M-Sport	-	-
036	Boreham/MLP	-	-
037	M-Sport	R5 FMC	1998 Works Valvoline car
038	M-Sport	R6 FMC	1998 Works Valvoline car
039	Boreham	-	-
040	M-Sport	S13 FMC	1998 Works Valvoline car
041	[Not completed]	-	-
042	Boreham	-	-

Forty different works or works-blessed Escort WRC models were completed between 1996 and 1998. In addition, several other Escort WRC models were created by private owners in later years, for their own use, by updating Escort RS Cosworths with engine/rear suspension body kits provided by M-Sport. To complicate matters further, the Escort WRC 'replica' which featured on the Ford press fleet, was a 'lookalike' with a full-spec engine and aerodynamic kit, but it lacked major items of the WRC kit (for instance, it used a five-speed gearbox).

Except for Chassis Number 038 (R6 FMC), which was driven by Petter Solberg on the Swedish Rally of 1999, all the works cars were sold off or retired at the end of the 1998 season.

Rally car development and improvements

During 1991, and still well before the road car went into production at Karmann in Germany, the Escort RS Cosworth rally car development continued and intensified. There were occasional rally appearances in Britain where the pace of Malcolm Wilson's car offered great hope, and also in Spain where Mia Bardolet's car (the programme overseen by Mike Taylor Developments) won several events in the Spanish Gravel series where homologation was not required.

As far as prototype Escort RS Cosworth rally cars were concerned, it was the same story again in 1992. These machines used the very best of everything that had already been tried, tested and proven in the works Sierra Cosworth 4x4s, including the 300bhp-plus YB engines (some said that the real figure was 360bhp and more, but since the FIA was trying to limit power outputs, Ford and other rival manufacturers were always careful not to claim any more than 300bhp at the time), the seven-speed non-synchromesh gearboxes from FF Developments of Coventry, and the larger and more solid front and rear axles (with a 9in diameter crown wheel at the rear and an 8.5in diameter crown wheel at the front).

Magnesium wheels, huge brake discs, massive brake calipers, Bilstein struts and dampers, different front cross-members, different rear cross-members, bag fuel tanks and more and more and more, were all finalised. In the 1990s more than ever before, well-specified Group A rally cars could be very special indeed.

By 1993, the only Escort RS Cosworth rally car item still shared with the road cars was the bodyshell, and even then it was advisable for serious competitors to purchase a special assembly, complete with welded-in multi-point roll cage and thin gauge skin panels.

There was more. By this time, the works team and its associates had built up so much experience with the new car that they evolved two completely different packages – one for its own 1993 works team and another for the scores of private owners. The difference between the chassis types was profound. For the private owners, who might want

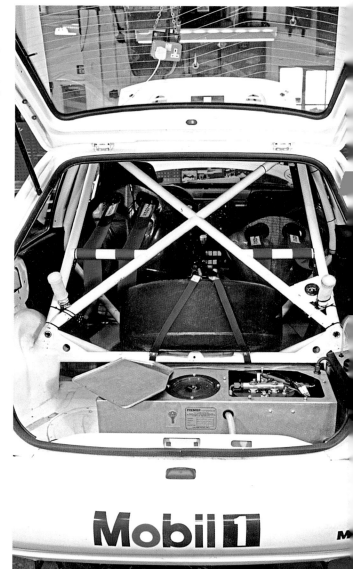

The original works Group A Escort RS Cosworths incorporated this complex roll cage, while the Premier bag fuel tank and the fuel pumps were all contained in the light-alloy box behind the line of the rear axle.

Like the Sierra Cosworth 4x4 which it supplanted, the Sierra Cosworth 4x4 engine was placed 'north-south' (in line),

This was the instrument panel layout of a 1993-model works Escort RS Cosworth. All the team cars had left-hand drive.

to create a new Escort RS Cosworth rally car by buying a shell from Boreham, and using almost all the running gear from their existing Sierra Cosworth 4x4 competition cars, new pieces were often interchangeable from those of the 1990-1992 works Sierras. In 1993, scores of 'new' Escort RS Cosworth race and rally cars were created like that – in fact, Boreham's Motorsport Parts department started delivering bare, white-painted bodyshells before deliveries of road cars

even began. For the works team, though, John Wheeler and Philip Dunabin's test teams had established that there was a need for more suspension wheel movement to improve the traction, so for works use only, there was a new and integrated wide-track package, with different springs, struts, dampers, track control and wishbone arms, anti-roll bar kits, magnesium semi-trailing arms, drive shafts and steering gear. However, it wasn't possible to cherry pick one or two items from this little lot – to make everything work it was either all or nothing, as the less well-financed private owners soon discovered!

This was a period in which Ford's personnel changed completely, and rapidly. Director, Stuart Turner, retired at the end of 1990. Peter Ashcroft succeeded him, but retired at the end of 1991, and his successor, Colin Dobinson, was experienced at Ford but an outsider to the world of rallying, and was known as a formidably efficient marketing personality and business manager. At the same time, in 1990, Chief Rally Engineer John Wheeler moved away from Boreham to Dunton, to usher the Escort RS Cosworth road car into production. His place at Boreham was taken by his one-time deputy, Philip Dunabin. Ex-driver John Taylor became the rally service supremo – and would soon move further up the scale, for in 1995, after Peter Gilitzer had become the Motorsport Director, he became Operations Manager, and effectively ran the team on all the events.

In 1994, the Escort RS Cosworth's second World Championship season, the team's most senior engineers, John Wheeler and Philip Dunabin, chipped away at making

The massively strong Group A front suspension uprights and big ventilated disc brakes were part of the rally car specification in 1992/1993. As time went on, the brakes would be enlarged even further.

further improvements to the rally car's specification. Even wider-track suspension was used on tarmac, and later on gravel-specification cars, anti-lag fuel injection systems were tried out on the exhaust side of the engine (this showed itself off by the machine gun-like back fires which this car then used to emit on the over-run), and there was much detail work on modifications to the integral bodyshell roll cage layout.

There was continual work on the engine by Mountune, the team's engine builder, though the compulsory restrictor fitted ahead of the turbocharger inlet meant that no extra peak horsepower, only extra peak mid-range torque, could ever be found. Much experimentation went on in transmission differential settings, while a sequential gearchange system was tried, but later abandoned, and a six-speed, rather than the original seven-speed, version of the main gearbox was built and tested, but would not be fully adopted until 1995.

Technically, therefore, the main innovations in 1995 concerned the transmission. Work on the sequential gearchange, which had been trialled in 1993 and 1994, was quietly abandoned, and from Corsica onwards, the team took to using the six-speed transmission. Traction control

37

This special light-alloy, semi-trailing arm formed part of the rear suspension of a Group A Escort RS Cosworth. Note that the length and attitude of this assembly could be adjusted at the front (inner and outer) pivots. Standard cars used a pressed steel item.

Mountune, as ever, worked away at the engines, within ever-tightening FIA regulations. Six-speed transmissions were standardised (except on the Safari, where seven-speed boxes were preferred) modified semi-trailing arms were developed in the rear suspension, and Dynamic shock absorbers, which had remote reservoirs, were also used. This was the first time for many years that Ford and Boreham had not preferred Bilstein dampers.

By mid-season, however, every effort was being concentrated on the design of the new Escort WRC, which began its career in 1997. The early works cars used Dynamic dampers, but later changed over to Hi-Tech items instead, and during the season, an Xtrac gearbox offering a sequential change, was also adopted.

Due to the fact that M-Sport, on behalf of Ford, started development of the all-new Focus WRC in the spring of 1998 – it was to make its debut in Monte Carlo in January 1999 – there was neither time, nor much inclination, to make any sweeping changes to the Escort WRC in 1998, though the works cars were now light enough to be able to carry ballast to meet minimum-weight restrictions, and launch control became part of the transmission.

Building and running the works cars

Boreham had been Ford's centre of rallying activities since 1963, using a workshop complex on the edge of an ex-USAF airfield north-east of Chelmsford, the runways, perimeter track and dispersal areas of which had been used for Ford testing since the 1950s. Changes were continually being made to update the premises, and allow for the increasingly high-tech nature of the cars prepared therein, although work on the Escort RS Cosworth was simplified by the fact that much of the chassis of the car had evolved from that of the Sierra Cosworth 4x4s of 1990-1992.

However, in the winter of 1992-1993, a series of big changes took place in the layout of Boreham's main workshop floor. Of course, there was never likely to be enough space - it seemed to have been like this since 1963 when the new building opened. During the Sierra period, cars had somehow been created by using both sides of a

and 'active' centre differentials were both adopted during the season.

Like all other teams, Ford had converted to using a turbo anti-lag strategy, and to meet the latest regulations, it also had to adapt to using a compulsory, smaller, turbocharger restrictor of only 34mm diameter, along with revised engine details. In an interview published years later, Philip Dunabin admitted that for 1995 the actual peak horsepower had therefore been reduced by 50-60bhp. Since he also said that the 1995 engines gave about 300bhp, this meant that the 1993/1994 cars had been good for at least 350bhp, if not more. So much for the FIA's wish that Group A rally cars should have no more than 300bhp! Philip also confirmed that the choice of a six-speed, as opposed to a seven-speed, transmission, was a direct consequence of the change in engine power and torque characteristics.

As Boreham's management team – Martin Whitaker from mid-season, Philip Dunabin and John Taylor across the whole period – became bound up in the rapid evolution of the Escort World Rally Car, the existing RS Cosworth specification was stable for much of the 1996 season.

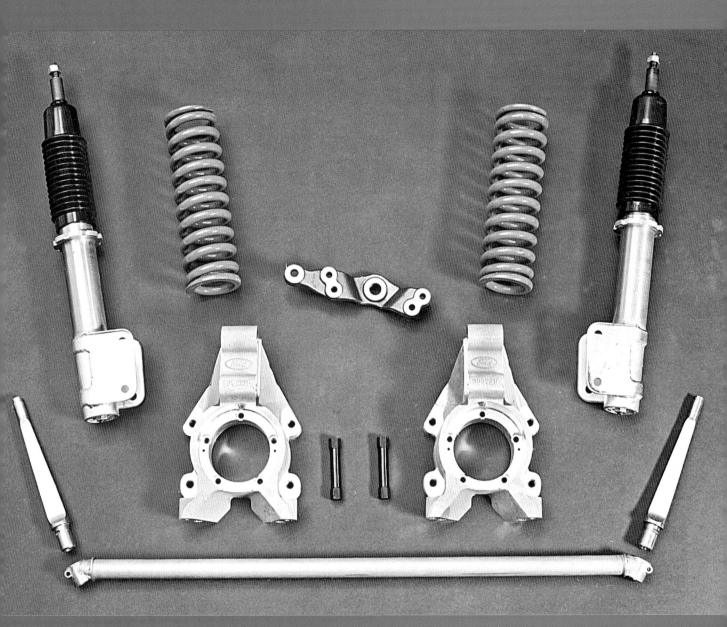
As posed for homologation pictures needed to gain approval for the Group A Escort RS Cosworth in late 1992, these are major components of the front suspension. The struts are adjustable for height, and different stiffness anti-roll bars can also be fitted.

central gangway, when up to ten bodyshells had jostled for attention. In 1993, all this changed. The floor space was re-jigged to provide just six work bays. To relieve the overcrowding, Gordon Spooner Engineering (in the nearby town of Witham) took on the building and maintenance of all the Group N recce cars. At Boreham, small teams of mechanics concentrated on building cars for each driver, and there was a nice touch in that both Miki Biasion and Francois Delecour were provided with specially tailored Sparco seats with their names embroidered on the backrests.

Success was now expected, the staff at Boreham could make no excuses – they were, after all, about to go rallying with a car they had conceived for themselves, partly engineered for themselves, and had spent two years testing and occasionally rallying.

For a short while, there was discussion about running three cars, though this was never done on a regular basis. From time to time, there would be three works Escorts on an event, though one of them would often carry its own sponsorship and be detached from the Boreham cars in some ways. By now, at least, there was no question of split tyre contracts – for both Biasion and Delecour ran on Michelin rubber.

"Our expectations were much higher. Here was a brand new car," Dobinson admits, "which had apparently overcome all the deficiencies of the Sierra. It was lighter, smaller, and more nimble – all that stuff. When we went to Monte Carlo, everyone was waiting with bated breath to see how it performed. Only then did we say to ourselves that here, finally, was a car that could do very well. We were never uncompetitive in 1993. But we certainly didn't kid ourselves that it could win every event."

In 1993 and the early months of 1994, the new Escort RS Cosworth enjoyed great success. As usual, however, with a Ford Motorsport programme, the financial and marketing interests within the company were heavily involved, and the fact that Francois Delecour was badly injured in a non-rallying road crash in April didn't help, especially as the substitute drivers who had to be employed while he recovered did not seem to be able to win.

By the middle of the season, Ford Motorsport at Boreham appeared to be in continuous turmoil. In July 1994, Motorsport Director, Colin Dobinson, suddenly announced that Ford's works team would close down at the end of the year, and that Boreham would become a mere research and development (R&D) centre for prospective Motorsport products, in particular to look at future RS2000-based F2 projects. This, we now know, was mainly hot air, though it was certainly true that there was not only a crisis in funding, but in top-management resolve to remain in the sport without outside help. At that precise moment, too, we did not know that Dobinson himself was soon to move on, leaving the company which he had served for so many years.

The immediate result of this shock announcement was that Ford began looking around to find other contractors to take over the running of cars in the 1995 World Rally Championship, instead of keeping the operation at Boreham open. Two teams in particular – RAS in Belgium, and Schmidt in Germany, both of whom had been running competitive Escort RS Cosworths – made bids for this contract. At this stage, there was no move from Malcolm Wilson's little company, indeed M-Sport, as it later became known, had not yet even come into existence.

To make the proposition financially viable, Ford made it known that it expected that teams would have to guarantee an absolute minimum of £200,000 to put two cars on the start line of an event. Even if all events were not tackled, therefore, Ford was looking for a commitment of at least £2 million.

Shortly after this, Colin Dobinson made his second shock announcement – he was going to take early retirement – which meant that Motorsport would soon have its fourth boss in five years, a statistic that could not have been desirable for the long-term stability of the operation.

Dobinson's replacement was to be 56-year-old, Peter Gillitzer, a Ford Australia motorsport personality who had only recently arrived in Europe, and was clearly a favourite son of Ford of Europe's CEO, Jac Nasser. Gillitzer took over the operation at once, and for the next few months found himself faced not only with Boreham's turmoil, but a severe lack of budgeted finance with which to fund the 1995 season. Not only was the works team due to be closed down, but at one point it also looked as if the successful British Touring Car Championship programme, run by Andy Rouse Engineering with Mondeos, would also be chopped. Gillitzer was a stranger to most people in British motorsport circles. He was not yet well known within Ford, either, and he might not have known much about motorsport – there were some personalities at Boreham who never forgave him for that – but he was a blunt straight-talker in the Ford corridors of power, and he proved to be a master tactician.

At the very end of 1994, he not only mapped out Ford's rallying future in a novel way, but he also saved the BTCC/Mondeo programme as well. History now tells us that he should perhaps have spent all the funds of those programmes on only one activity, and made sure that the chosen activity was operated really well, but at the time, it was a brave and dextrous move.

Certainly the closure announcements of mid-1994 were firmly swept away. Later in the year, it was declared that Boreham would remain open for 1995, and that the works Escort RS Cosworths would continue to be developed and prepared in those historic workshops, though RAS Sport would sometimes be preparing Bruno Thiry's car in Belgium. The cars were to be entered by a new team to be called RAS-Ford.

This was the moment at which the very important personality, John Taylor, became Operations Manager. RAS Sport, it was stated, would run the cars in events managed by its own Vittorio Reisoli. However, this arrangement did not seem to last long, and Taylor was established as the figurehead. With the works Mobil sponsorship deal now coming to an end at the close of 1994, much of the sponsorship for the new cars would come from Belgium – notably from Giesse (an automation company), Fina (petrol and lubricants), and Bastos (a Belgian brand of cigarettes).

There was more management upheaval. Top engineer, John Wheeler, chose this moment to move on to greater things – he became Chief Engineer on the Aston Martin DB7 project in February 1995 – which meant promotion at Boreham for Philip Dunabin, who would stay in post until Boreham closed down completely, in 2004.

So who, or what, was RAS-Ford? RAS was a Belgium-based company which had started out in saloon car racing in 1983, had then moved up to running Robert Droogmans' successful Group B Ford RS200 rally car in 1986, and then joined forces with Astra of Italy, to run several successful Ford and Lancia rally cars in the early 1990s. Team boss Mauro Pregliasco, himself a one-time Lancia works rally driver, certainly had the experience and his team the background to turn this into a success.

Yet it was not. In a season when the Escort RS Cosworth should have been at its peak – independent engineers still consider that it had the ideal layout for a modern, powerful and versatile Group A four-wheel drive car – the team often struggled, not only for pace, but for credibility. The rallying media soon noted that RAS-Ford personalities clashed repeatedly with Ford Motorsport at Boreham, the driving team was neither world class nor consistent, and, at times, there seemed to be a distinct lack of enthusiasm among the personnel. A few years earlier, with strong, window-rattling, managers like Stuart Turner or Peter Ashcroft at the helm, it might all have ended well, but both of these towering personalities had retired, and there seemed to be a lack of direction in their place.

The short-lived joint project with RAS came to an end after only one year, the split being announced before the end of the 1995 season. With the strong, not to say, bullish, characters of John Taylor and Philip Dunabin reporting to Peter Gillitzer, it could so easily have been impossible for the group at Boreham to carry on. Fortunately, and dramatically, however, the signing of twice World Champion, Carlos Sainz, to lead them into 1996, quite transformed the atmosphere, and the World Championship team carried on successfully.

Soon, however, another upheaval took place. For 1997, and with a major new derivative, the Escort World Rally Car, due to start rallying, Ford decided to contract out its rallying business for the very first time. Except in the years when David Sutton's team had been running the Eaton's Yale and Rothmans Escorts, all works or quasi-works Escorts had

Malcolm Wilson began his rallying career as an impecunious enthusiast. He rose to drive for the Rothmans Escort team in 1981, built up his own businesses, won the British Rally Championship in an Escort RS Cosworth in 1994, and then founded M-Sport in 1996 to run the World Championship rally team.

been based at Boreham. Now, at short notice, the newly-installed Director, Martin Whitaker, approved a plan that would turn Boreham into an R&D centre, while Malcolm Wilson's nascent M-Sport operation would build and run the works rally cars themselves.

Geographically, this made little sense, as Malcolm's business was based in his native Cumbria, close to Cockermouth which was well over 300 miles or almost a hard day's drive north of Boreham. However, there was no doubting Malcolm's enthusiasm, nor the rapidly assembled expertise of his organisation. Officially formed on 18 November 1996, even with a small but efficient little workshop ready to start work, few expected this to be a success before mid-season. But it was! Just two months later two new cars took the start in Monte Carlo. Both finished and Carlos Sainz came within 55 seconds of outright victory.

By November, homologation of the new Escort WRC was already in sight, and achieved on 1 January 1997 with a mass of components either available or on order. Apart from the two prototypes, however, no other complete cars were ready, or even being assembled. As Malcolm Wilson commented soon afterwards: "Our priority was to get the cars to a condition where the drivers could just step straight into them. The problems were many. We had to build cars of a type that had never been rallied before, cars nobody knew anything about. We had to have parts produced that had never been made before. We had to recruit a workforce that would work together well in the future ... And all the time we had to prepare for the future."

This, then, was the new face of the Ford works team, where the building and running of cars, rather than hosting the press and providing glossy brochures, was all-important. Though John Taylor immediately left the company, Boreham did not close its doors, but carried on building customer Escort WRCs, as well as selling millions of pounds' worth of components to private teams all around the world.

Personalities and star drivers
Stuart Turner

Although Turner's best-selling autobiography is entitled *Twice Lucky*, there was no luck about it. Already famous as BMC's Competitions Manager in the 1960s, along with Walter Hayes he was one of Ford's most influential Motorsport characters from 1969 to 1990. Turner initially trained as an accountant, but broke into rallying at club level, after which progress as a club magazine editor, a works co-driver and the first Rallies Editor of *Motoring News* was logical.

Having won three national Championship co-drivers' awards, and sitting alongside Erik Carlsson when he won the RAC Rally of 1960, he then became BMC's Competition Manager from 1961 to 1967, when the Mini was at the height of its powers. Headhunted by Walter Hayes to Ford Motorsport in 1969, where he succeeded Henry Taylor, his ruthless methods and sure eye for publicity stamped his authority on everything which Ford tackled in rallying.

After becoming Director of Motorsport in 1970, he then managed the AVO plant from 1972 until it closed down in 1975. For the next seven years he directed the fortunes of the Public Affairs department, before returning as Director of Motorsport, Ford of Europe, in 1983. In 21 years at Ford, he was a prime mover in the Escort RS1600/Mexico/2000 programme, in the Mk 2 versions of those cars, in the Escort RS Turbo, the RS200, the Sierra RS (and RS500) Cosworth, and, finally, in the Escort RS Cosworth. On the other hand, it was Turner's decisive axe which killed off the RS1700T.

Retiring from Ford at the end of 1990, he turned to his next career – as a superb, and much-in-demand, after-dinner and conference speaker, and was still enjoying this 'hobby' as the 2000s began.

Mike Moreton

As a planner, organiser and schemer, Mike Moreton had no equal. In print, I once defined Mike as a 'fixer', and, though he thought this was a touch undignified, he never actually complained – I reckon he was proud of that description. Whether it was planning new AVO products, or aiding and abetting Stuart Turner and John Wheeler to get the Escort RS Cosworth approved, he was peerless.

After working at Rootes in Coventry, and at Vauxhall, Mike joined Ford in 1966. Within two years he had hitched his star to Bob Howe, and became AVO's product planner in 1972. After a period in mainstream Product Planning, where he wrote an influential paper suggesting that a Special Vehicle Engineering department be set up (it was!) he was then seconded to Boreham to make planning sense of the Escort RS1700T in 1982, though that work, in the end, was wasted. He was an important mover and shaker at Boreham for some years, heavily involved in the development of the Escort RS Turbo and the Sierra RS Cosworth, but he then became Project Manager of both the Sierra RS500 Cosworth and the demanding RS200 projects.

In 1988 and 1989 he was a founding father of the Escort RS Cosworth project – with Engineer, John Wheeler, and Director, Stuart Turner – pushing it over every product planning hurdle, and finding a production home for it at Karmann in Germany. He was then headhunted by Tom Walkinshaw, to work at TWR in Oxfordshire, and his direct Ford connection was broken. In a decade at TWR he was Project Manager for the Jaguar XJ220, ran the Aston Martin factory at Newport Pagnell for a time, inspired the planning behind the Aston Martin DB7, and advised Walkinshaw on many other projects.

Peter Ashcroft

'Mr Nice Guy,' one of Ford's most famous personalities and respected by everyone, Peter Ashcroft was involved in the development, or rallying, of almost every Escort RS model.

Originally a mechanic with the Gilbey Engineering racing team, he joined Ford Motorsport in 1962. Except for a short period when he worked with Brian Hart at the Peter Sellers racing team, Boreham was his workplace, from 1963 to the end of 1991. Initially working as an engineer, and later running the engine development and test department at Boreham, he spent two winters at Ford Cologne (where he inspired the improvement in Capri RS2600 engines which turned those cars into Touring Car winners) before becoming Ford UK's Competition Manager in 1972. He ran the development and build programmes for the works Escort Twin-Cams and RS1600s, as well as shaping the 1.8-litre engines, which used the 'Ashcroft' block, and developing the 140bhp London-Mexico engines of 1970.

From 1972 until the end of 1991, he led the victorious works Escorts teams and guided the department through the confusion of the RS1700T, the front-drive RS1600i, the RS200 and the rear-drive Sierra RS Cosworth eras, before being closely involved in the original concept of the ACE, the Escort RS Cosworth project.

In many ways, it was Peter's strategy which helped his successor, Colin Dobinson, to enjoy such a successful tenure of office, in 1993 and 1994.

Colin Dobinson

When Colin was appointed as Ford's Director of Motorsport to replace Peter Ashcroft, he was almost totally unknown outside the company. No matter. With Ford he already had a formidable reputation in marketing, and took no time at all coming to terms with the new position. It was under Dobinson's calm and always logical management that the Sierra RS Cosworth became a nearly great rally car in 1992, and it was no coincidence that he was running Boreham when the Escort RS Cosworth hit the ground running, and notched up six World Championship victories which climaxed in Francois Delecour's famous win in Monte Carlo, in January 1994.

Unhappily for him, though, he then had to preside over what he once described as an 'agonising re-appraisal' of Ford's future in motorsport. He was virtually bulldozed into accepting a scheme whereby Boreham would become no more than an R&D centre, and the actual rallying would be farmed out to a contractor. Because his antennae were acutely sensitive, he realised that this would not be popular,

Peter Ashcroft – 'Mr Nice Guy' as he was known by everyone – ran the Boreham complex for many years, and retired when he had ensured that the Escort RS Cosworth would go into production and be rallying soon.

Team boss Colin Dobinson (right) and driver Miki Biasion, wonder how they can get around the dust cloud problems of Rally Argentina in 1993.

so when the opportunity arose to take early retirement, and become a director of an engineering consultancy, he speedily took it, leaving Boreham late in 1994.

John Wheeler

London-born John Wheeler was always interested in automobile engineering, and came to Ford almost by chance in 1980, after spending years with Porsche. He was a rising star in the chassis area at Porsche (this including work on racing sports cars) when in 1980 he answered an *Autosport* advert – for a job at Boreham.

Once there, he soon became recognised as an absolute expert on all chassis design and behaviour matters. He led the team which designed the still-born Escort RS1700T, and lobbied in vain for a four-wheel drive version to be developed. Later, it was John's concept for the RS200 which was combined with the more race-track orientated approach of Tony Southgate, their joint efforts evolving into the 200-off super-car. After this he was appointed Chief Engineer on the rally-improvement of Sierra RS Cosworth cars. From 1988, he was always closely involved in four-wheel drive and transmission possibilities. Egged on by Stuart Turner's "Why don't we ...?" approach, he partly conceived, then schemed out, then analysed the possibilities, and then ran the initial engineering programme behind the new ACE (Escort RS Cosworth) project.

A spell as Aston Martin's Chief Engineer then led to his return to Ford's technical HQ at Dunton, where he worked on a variety of secret projects.

John Taylor

Born into a naval family, initially a very successful steeplechase jockey who turned to rallycross when back injuries made further horse racing potentially perilous, J.T. (as he was always known) became European Rallycross Champion in 1973. Having forged strong links with Ford RS Dealer, David Haynes of Maidstone, he then built a successful rally career in Escorts. He effectively became a member of Boreham's B Team, but retired in 1979 when his recurrent back problems returned.

Two of Boreham's most important personalities at this time were engineer John Wheeler and ace-driver Ari Vatanen.

In the 1980s he blossomed at Boreham as teacher, guru, and organiser, eventually taking on the complete re-organisation and re-equipment of the service set-up. Stuart Turner and his successors all believed in his forthright views – they were right, of course – and it was J.T.'s perceptive eyes which discovered drivers like Carlos Sainz, Mark Lovell and Louise Aitken-Walker. As the 1980s and 1990s progressed, he became an essential sounding board, to be consulted on all Ford rallying matters.

Under Colin Dobinson, and even more so under the short-lived Director, Peter Gillitzer, as Operations Manager he virtually ran the Ford rally team on the ground. Not only did he personally persuade Carlos Sainz to return to Ford for 1996, he also inspired the rebirth of the team for the season which followed. When Ford decided to contract out the rally operation to M-Sport at the end of 1996, J.T. was instantly made redundant – as big a travesty of justice as could ever be imagined.

Malcolm Wilson

Cumbrian-born Malcolm started driving scruffy old Fords well before he had a driving licence, and the family scrap yard business made it easier for him to afford a fast Escort when he did. Speedily moving up from Mk I to Mk II, he joined the Rothmans rally team in 1981, drove works Escort RS1600is in 1983, spent time testing the new RS200, rallied his own (ex-Sutton) Audi Quattro, then drove for Austin-Rover, Vauxhall and Peugeot, before returning to Ford with the Sierra Cosworth 4x4 in 1990.

Apart from spending many long hours as a Motorsport test driver, he also tested, developed, and rallied the new Escort RS Cosworths, finally winning the 1994 British Rally Championship in his Michelin-sponsored car. In the meantime, Malcolm Wilson Motorsport (the predecessor of M-Sport) expanded to prepare cars for other teams, particularly in Europe. So when Boreham contracted out its works rally operations in 1997, Malcolm's business was an obvious candidate.

Not only did M-Sport then campaign the Escort World Car with great honour in 1997 and 1998, it also engineered the all-new Focus WRC, and persuaded Colin McRae to lead the team. In more recent years, M-Sport, the Focus and Ford's own efforts became closer and ever-closer together, their crowning glory being to win the World Makes Championship twice, in 2006 and 2007.

John Taylor was closely involved with the Ford works team from 1974 to 1996. He became Operations Manager in 1995 and 1996.

Immediately before leaving to start their 1993 rally season, Francois Delecour (left) and Miki Biasion pose at Boreham beside Francois's Escort RS Cosworth.

Francois Delecour

Except to a limited number of his native Frenchmen, Francois Delecour was almost unknown when Ford signed him up for 1991. Yet he almost won his very first event, in Monte Carlo, until he was foiled by a suspension breakage, and was always the most effective of Sierra Cosworth 4x4 drivers thereafter.

Taking to the Escort RS Cosworth like a duck to water in 1993, Francois was second in Monte Carlo, but thought he deserved victory because of Toyota's questionable tactics. He was then victorious in Portugal and Corsica, before winning another World event in Spain, and he always looked likely to win on every other round. His biggest victory, unquestionably, came in Monte Carlo in 1994, but his career was then blown off course after he suffered broken ankles in a non-rallying road crash, and he never quite recovered his flair, nor recorded more victories, after that.

Released by Ford early in 1996, Delecour then drove works Peugeot F2 cars for a time, before coming back to drive the Ford Focus WRC in competent, if not outstanding, fashion. In many ways a fiery, not to say explosive character, he inspired great enthusiasm among many rally fans, but was not always easy for a team boss to manage. He will, however, always be recognised as the most charismatic character in the Escort RS Cosworth's orbit in those early years.

Miki Biasion

Is it cruel to suggest that good Italian drivers, like good Italian wine, don't seem to travel well? When Miki Biasion joined the Ford works team for 1992, he had already been World Champion twice (1988 and 1989), and had come to seem like a fixture in the Lancia works team. When Peter Ashcroft captured him, on a three year contract, it was seen

Mike Biasion (left) and Tiziano Siviero were the Italian team which brought so much experience to the Escort RS Cosworth programme in 1993 and 1994.

as a real coup, but the successes never quite rolled out as hoped. Although Miki was always professional, and always diligent, he would only win one event for Boreham, and was even seen to be over-shadowed by the sheer exuberance of Francois Delecour.

Having leapt to fame by becoming European Rally Champion in 1983, he then remained faithful to Lancia until 1991, becoming the master of the Delta Integrale, and recording no fewer than 16 individual victories. So, should Ford have been more suspicious when Lancia didn't fight too hard to retain him after 1991? After spending 1992 making sense of the recalcitrant Sierra Cosworth 4x4, Biaison then started the Escort RS Cosworth's career, in 1993, with one victory, two second places and a third. Then came the disappointments of 1994, for although he became de facto team leader after Delecour's road accident, he did not win another event. He seemed to become disillusioned with the car, and the team itself, and faded away. Having left Boreham, he did not then drive for any other works team, and soon dropped out of the sport altogether.

Tommi Makinen

As far as Ford is concerned, Tommi Makinen was the future rallying superstar who got away. Like other Finns who went on to become famous, Tommi started rallying in an Escort RS2000, in 1985, but his early World entries were in cars as various as a Lancia Delta HF 4x4, a Mitsubishi Galant VR4 and a Nissan Pulsar GTI-R. Awarded a one-off works drive in an Escort RS Cosworth – the 1000 Lakes 1994 – he recorded an emphatic victory, his very first.

And that was that – for Tommi never again drove a Ford at World level. Instead, he signed up for Mitsubishi, with whom he won the World Drivers' Championship on four consecutive occasions. With all due respect to Carlos Sainz and Juha Kankkunen, what a great loss this was to Ford.

Ari Vatanen

If Hollywood was ever to design a template for a world rallying superstar, the result would be Ari Vatanen. Handsome, friendly, charismatic, possessed of unearthly driving talents, and with the will to win in any car, anywhere, Ari was every enthusiast's hero.

Recruited by Ford in 1975, he drove nothing but rear-drive Escorts until 1982, became World Champion in 1981, then went off to drive for Peugeot, where an awful accident in Argentina nearly killed him in 1985. He won Paris-Dakar several times and then drove for Mitsubishi teams, before semi-retiring and devoting himself to European politics, becoming a Member of the European Parliament.

Fortunately available in the mid-1990s, when Ford's Francois Delecour broke his ankles, Ari put up several spirited performances in works Escort RS Cosworths, before finally turning away from active motorsport.

Although he did not win a World event for the Ford works team, he is the personality and the name that many Ford fanatics link to the Escort.

Carlos Sainz

No-one ever had a bad word to say about Carlos, the multi-talented Spaniard who was well connected in Spanish court circles, and could equally have made a sporting

Carlos Sainz (left) and Malcolm Wilson were both important members of the Ford works team in the 1990s.

career in football, squash or tennis. First discovered by Ford in 1987, he moved on to Toyota, Lancia and Subaru before re-joining Ford for 1996 and 1997. Along the way, he won the World Drivers' Championship twice, and in those two seasons with Ford he recorded three victories and no fewer than seven seconds and three third places. Almost single-handedly it seemed he revived Ford's morale in 1996, having a self-belief and a faith in the cars which Operations Manager John Taylor had said was there for the taking.

Not only supreme and very consistent on tarmac, loose surfaces, or ice and snow, Carlos was one of the best and most dedicated test drivers in the business. Cultured, polite and helpful to everyone he met, he was liked by all, loved by most of the Spanish nation and an inspiration to any rally team. Having left Ford after the Escort WRC's first season, he would return to drive the Focus WRC in 2001 and 2002, and by the time he finally retired in 2004, had won no fewer than 26 World rallies.

Juha Kankkunen

Although he started rallying in Finland in 1978 in his own Escort RS2000, Juha Kankunnen did not join Ford's works team until mid-1996. His first works drives were with Toyota in the mid-1980s, and in a whirlwind career he then drove Peugeots, Lancias, Toyotas (again) Lancias (again) and Toyotas (again) before finding himself out of work in 1996 after Toyota was thrown out of World rallying for cheating over engine/turbocharger restrictor infringements.

Joining Ford from Argentina in 1997, Kankkunen then stayed with the team until the Escort WRC was finally

In 1998, Juha Kankkunen (left) and Bruno Thiry were the works team's high-profile drivers in Escort WRCs.

sidelined at the end of 1998, when he refused to play second fiddle to Colin McRae in the new Focus WRC. After this, he ended his rallying career with appearances in Subaru Imprezas. Not known as one of rallying's deep thinkers or communicators, he seemed happier on loose surfaces and a golf course than on tarmac and in press conferences. He was, nevertheless, consistently fast, very rarely crashed his cars and was a great team member, especially in 1996 when he had to give way to Carlos Sainz in the Ford team.

Although he was World Rally Champion no fewer than four times in the 1980s and 1990s, and recorded 23 individual rally victories, Kankkunen never quite managed to win for Ford. In two seasons, however, he recorded seven second places and five thirds.

Competition story

After much high-pressure development work, the Escort RS Cosworth was previewed in 1990 and went on sale in the spring of 1992. Features standard on all production cars included the massive wheel-arch extensions, the 8in-wide wheel rims, the big front spoiler, and the louvres in the bonnet panel.

'First Time Out, First.' That was Ford's advertising slogan after the very first Escort RS Cosworth rally car won the Spanish Talavera event in October 1990.

It was an ideal and novel way to get a new rally car's career under way and, as far as I know, no other serious car manufacturer ever took the same approach. The Escort RS Cosworth was a rally car which went into production to gain homologation, not a road car that just happened to be a good racing car too. This was just one reason why the first rally car appeared nearly two years before road cars went on sale – and why it was an immediate winner.

When the new-generation Escort was approaching launch, in September 1990, and with the Escort RS Cosworth still a closely guarded secret, Stuart Turner realised that any pizzazz he could offer would help. The unexpected preview at Blenheim Palace has already been mentioned, but this was only the start: "... there was an ideal opportunity to run our first prototype rally car in September 1990," Turner recalls, "in the one day Talavera Rally in Spain, which had no homologation restrictions ... If we took the car out there, with one of our contracted teams – Mike Taylor Developments, with John Taylor running the team – and it failed, I reasoned that we could ignore the whole episode,

Although it was still almost totally undeveloped, this Escort RS Cosworth was good enough for Mia Bardolet to drive it to victory in the Spanish Talavera Rally of 1990.

and hope that no one had noticed. But if we succeeded – well, we could make as much noise as if we had won the Monte Carlo Rally."

The rest is history. Gordon Spooner Engineering built the car, Mia Bardolet drove it to victory, and, as Turner also quotes: "Once the car had appeared in public, and a success ad had run saying 'First Time Out, First.' I knew that it would be almost impossible for the company to cancel the project without huge loss of face."

For the next two years, development of rally cars went side by side with that of the works Sierra Cosworth 4x4s, which shared much of their running gear. Far too much nonsense denigrating the Sierras has been written – there is no doubt that they were, almost, extremely successful, especially in the 1992 World Championship season, and they were certainly cruelly unlucky at times. This was the period, incidentally, when the works rally engines, complete with 40mm turbocharger restrictor, produced 340bhp at 7000rpm.

With these four-wheel drive Sierras, a suspension breakage on the last special stage of the 1991 Monte Carlo cost Frenchman, Francois Delecour, outright victory, while

Malcolm Wilson loved the Escort RS Cosworth from the moment he first drove it – always happy if he could get it as far sideways as he had a Mk II Escort in the 1970s!

In 1992, Delecour and the Italian, Miki Biasion, were on World Championship Rally podiums four times, and Ford took third place in the Manufacturers' Championship.

Four-wheel drive Sierras also won ten European Championship rallies, along with six National Championships in the same year – which should have been enough to silence the loud mouths, but did not always succeed.

Although his un-homologated Escort RS Cosworth could not compete head-on with other cars in the Scottish Rally of 1991, Malcolm Wilson drove in a special extra category, and established a series of remarkable stage times.

K57 EVX was an early Group A RS Cosworth which was used for testing and development in 1992-1993, leading up to the 1993 Monte Carlo Rally.

did, however, provide a wealth of experience on which the works team could draw.

During 1991, still well before the production car went on sale, Escort RS Cosworth development continued. The occasional rally appearances in Britain, where Malcolm Wilson's pace offered hope, were really extended test sessions, and they went to Spain, where Mia Bardolet won several events in the Spanish Gravel series.

It was the same story in 1992. Prototype Escort RS Cosworth rally cars used the very best of everything which had already been proven to work well in the works Sierra Cosworth 4x4s, including the 300bhp-plus YB engines. (Some said that the real figure was 360bhp and more, but since the FIA was trying to limit power outputs, Ford was saying nothing at the time.) The *seven*-speed non-synchromesh gearboxes, which FF Developments of Coventry was manufacturing, and the larger front and rear differentials, with 9in diameter crown wheel at the rear and 8.5in diameter crown wheel at the front, were also included.

Magnesium wheels, huge brake discs, massive brake calipers, Bilstein struts and dampers, different front cross-members, different rear cross-members, bag fuel tanks and more and more and more, were all finalised. In the 1990s, more than ever before, well-specified Group A rally cars could be very special indeed.

1993

Ford was well placed for 1993 as, with Group A homologation achieved, works cars could enter World Championship events from 1 January 1993. Back at Boreham, the team anticipated an intensive programme. Having had much time to plan, it elected to commit the Mobil-sponsored works Escort RS Cosworths to ten World events, while Malcolm Wilson and Robbie Head would tackle British events. Everyone's favourite Welshman Gwyndaf Evans also landed a Group N British programme. Malcolm also took on the role of the team's principal test driver, and spent many hours both on the tracks at Boreham and on sessions in the UK and overseas.

For the first time since 1987, in order to score World Championship points, new regulations made it compulsory for cars to appear on the other side of the world as well as in Europe, so Colin Dobinson and his co-ordinator Melvyn Hodgson elected to send works cars into the southern hemisphere. Complex regulations? Fourteen events were nominated as World Championship qualifiers, a team could nominate entries in ten of those rounds, but only eight scores could be counted, six scores could be made in European events but the other two had to come from far-flung rallies. Still with me? It could have been a lot worse, but the days of compulsory entry in all World events was still in the future.

Mobil, the oil company, that had supported the works Sierras so enthusiastically in 1992, stayed onboard as the team's principal sponsor, along with Michelin, which made Ford one of its most favoured clients. Cosworth advised about engine preparation, but all the actual engine building

Not originally a motorsport enthusiast but a quick-thinking, incisive and well-liked marketing specialist, Colin Dobinson ran the works Ford rally team from 1992 to 1994.

and development was carried out by Mountune in Maldon. FF provided the gearboxes and axles, and was involved in much forward development.

Colin Dobinson started his second season as Director, Melvyn Hodgson had settled down as Operations Manager, while John Wheeler and Philip Dunabin remained as the team's pivotal engineers. John Griffiths was the ever-present guru, specialising in homologation and engine development.

For 1993, changes in homologation rules all seemed to work in the new Escort RS Cosworth's favour. Minimum weights (without roll cages) for the turbocharged 2-litre class rose by 100kg/220lb to 1200kg/2646lb – and the Escorts could usually run down to that limit, even on rough events where extra strengthening was needed. Maximum wheel rim widths were reduced by one inch (the Escort's chassis was always tailored to take advantage of this), and a standard, FISA-specification fuel was imposed in place of the special

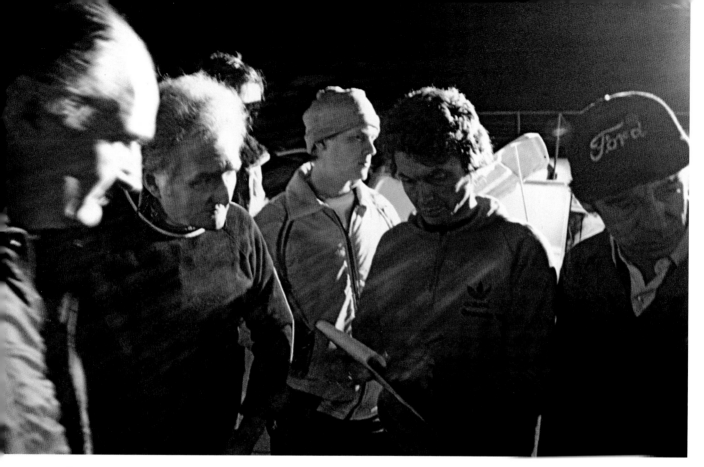

What problem was ever too difficult for these Boreham personalities to sort out? (Left to right) Peter Ashcroft, John Griffiths, John Wheeler and Mick Jones.

rocket fuel which top teams like Ford had been using in previous years.

Not only that, but to contain costs, in World rallying there was now a limitation on the number of turbocharger and major transmission swaps which could be made at service points during an event. Shorter hours became a feature (most events now included rest halts every night) which meant that events were getting shorter, sharper, and tougher.

Two brand-new rally cars were built for the 1993 Monte Carlo, and at the same time, a start was also made on other cars for Corsica and the Acropolis. Hampered by the new control fuel, the latest engine peaked at well over 6500rpm, and the agreed peak gear-change point was 7500rpm – not bad for a power unit which had been around since 1986.

The team's intensive season started in real controversy, became a triumphal progression around Europe, then hit misadventure in the southern hemisphere. Throughout the year, there was a straight fight between Ford and Toyota: we now know that although Toyota eventually took the crown, it spent at least three times more money than Ford could possibly commit.

Before World Rallying was rejigged to make central servicing more practical, Boreham had a big fleet of service vehicles to support the rally cars. Just before the team left for the start of the 1993 RAC Rally, three Escort RS Cosworths were supported by ten Iveco-Ford service vans, three trucks carrying bulky spares, tyres and wheels, a crew bus and seven 'supervision' ('chase') cars, too. Team boss Colin Dobinson about to retire after this event, is leaning against a rally car in the centre of the group.

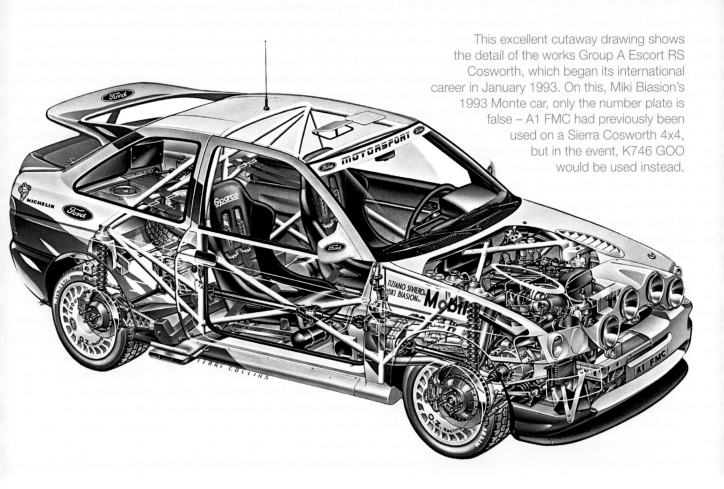

This excellent cutaway drawing shows the detail of the works Group A Escort RS Cosworth, which began its international career in January 1993. On this, Miki Biasion's 1993 Monte car, only the number plate is false – A1 FMC had previously been used on a Sierra Cosworth 4x4, but in the event, K746 GOO would be used instead.

For Monte Carlo the new cars (K746 GOO and K748 GOO) each had wide-track, long-travel suspension, and different transmission settings as specified by each driver. To support them, Boreham's back-up effort involved more than 100 people, manning up to 15 service vans, two trucks and a radio-communications aeroplane. In addition, there were ice-notes crews, management cars and supervision cars in profusion. This was Boreham's biggest ever effort – it was truly serious.

As in 1991, the first victory almost came at once. Until the penultimate special stage of the Monte, Ford looked certain to win on the Escort RS Cosworth's World Championship debut. Delecour, in particular, was devastatingly fast. Then Auriol's Toyota, soundly beaten by the Escort for the first three days, mounted an amazing charge on the final night and won by just 15 seconds. The way in which it occurred was controversial to say the least, and Toyota's victory was tarnished by suggestions of sharp practice, illegal fuel and illegal additives.

As *Autosport*'s rally report asked: "Was there any truth in stories of dubious fluids in the washer bottles allegedly plumbed to the engine via an extra injector? Does anyone

Ready to go at last! This was the compact Ford works team in Rheims, about to start the Monte Carlo Rally in January 1993.

really know what happened on those final stages? 'Toyota knows,' said Biasion."

Toyota, which was later thrown out of the sport for other, even more heinous, breaches, never objected to such published comments, so we may draw our own conclusions from that! Escort heroes, Delecour, officially second, and Miki Biasion, officially third, knew in their hearts that they had effectively notched up a 1-2. No other team – not

Mitsubishi, and not Lancia (Carlos Sainz was running a semi-official Jolly Club Delta Integrale in 1993) – could match their pace.

Boreham then ignored Sweden, but only five weeks after Monte Carlo, in Portugal, it made up for everything by recording the 1-2 success it thought it had deserved in Monte Carlo. Using the same, though re-prepared, Monte cars, Francois and Miki completely obliterated every other

Francois Delecour was the star of the Ford works rally team in the early 1990s. He drove Escort RS Cosworths to win the Portuguese, Corsican and Catalunyan (Spain) rallies in 1993.

Francois Delecour came within seconds of winning the Monte Carlo Rally in 1993, until a works Toyota rather mysteriously found some extras pace. No obvious rush at this service point, as mechanics (who were building the car at Boreham only two weeks earlier) check out everything before the next stage.

This we don't understand! Why are all six extra driving lamps uncovered in this 1993 Monte shot, but not a single one appears to be lit? Miki Biasion would take third in this event.

works team. Toyota did not appear (it had suffered further official wrath in Sweden due to illegal service support) and Colin McRae's Subaru Legacy was not quite quick enough. Delecour led almost from start to finish and set 18 fastest stage times, beating Biasion by just one minute but McRae by nearly three minutes. At Boreham, there was great rejoicing. This was the very first works victory since Didier Auriol won, in a Sierra RS Cosworth, in Corsica in 1988, the

Cold, but no snow or ice – this was typical of so much of the 1993 Monte, where the two works Escort RS Cosworths took second (this car) and third.

Francois Delecour and Daniel Grataloup finished second in the 1993 Monte Carlo Rally – and this very car would go on to win in Portugal just a few weeks later.

A famous victory – the first World success for the Escort RS Cosworth came in Portugal in March 1993, when Francois Delecour won in fine style, with team-mate Miki Biasion close behind him in second place.

first ever by an Escort RS Cosworth, and the first World Rally win by a Ford Escort since Ari Vatanen's Mk II Rothmans car had triumphed in Finland, in 1981. Amazingly, this was also the very first rally of any type that Francois had ever won outright! Congratulatory letters, faxes and messages poured into Boreham, though the team had no time to relax.

Would it be easy to achieve another victory? In Corsica? Well, Corsica was a different challenge.

Two more brand new, lightweight, tarmac cars – K831 HHJ (Biasion) and K832 HHJ (Delecour) – were completed. Their 17in diameter OZ road wheels hid 14in/355mm front brakes, with water-cooled callipers, so these cars steered,

The Escort RS Cosworth's first big win came in the Rally of Portugal, in March 1993, when Francois Delecour (right) and Daniel Grataloup drove this immaculate car. The RS Cosworth would win five World Championship rounds in 1993 alone.

Until he damaged his legs in a non-rallying crash in 1994, Francois Delecour was probably the world's fastest driver in an Escort RS Cosworth. Here he is, pushing on towards victory, in the 1993 Rally of Portugal.

Brand new for the occasion, Francois Delecour's tarmac-specification 'lightweight' Escort RS Cosworth won the Tour de Corse in fine style in May 1993.

handled and, in particular, stopped like no previous Fords had ever done.

The 24 stage, all-tarmac Tour de Corse became a head-to-head battle between Francois and Didier Auriol, though this time there were no dubious tactics from the Japanese team and the Escort led throughout: Delecour was fastest 15 times and Auriol only six times. For Biaison, a poor choice of tyres and a forced transmission change after a clutch failure led to road penalties, which didn't help, so he finished a rather dispirited seventh.

For Ford, this was a famous victory which almost, if not quite, made up for the disappointments of Monte Carlo – yet Boreham was too practical to look fondly on the winning car and put it into a glass case. After a few days' rest, it was used as a test car. For weeks, it kept its rally plates while management and the mechanics treated it very much like the family pet. Often parked casually outside the workshops when visitors came along (no, that sort of thing rarely happened by chance), driven by John Taylor, it finally established some remarkable performance figures for an *Autocar & Motor* report.

On dry tarmac using slicks, this hard-used Escort

When Francois Delecour's Escort RS Cosworth won the Tour de Corse in May 1993, it proved to be everything the planners had ever hoped for – that this new car could win on the loose and on tarmac against the world's fastest rally cars.

recorded 0-60mph from rest in 3.8 seconds, 0-100mph in a mere 9.9 seconds – in that time, J.T. had changed gear three times – and was on its way to a top speed quoted as 136mph. At the time, Mountune revealed (perhaps inadvertently, as FISA rather frowned on such power outputs) that the engine pulled 360bhp at 6800rpm, with 404lbft of torque at 4500rpm.

Boreham then prepared two more brand new cars for the Acropolis in May, these being heavyweights compared with those used earlier in the year, though they carried familiar-looking Monte/Portugal identities. Time, I am sure, to stop being so interested in registration numbers – especially as the author made a visit to Boreham at this time, and saw the way that brand of new car was built up.

Miki Biasion claimed a long-awaited victory for Ford when he won the Acropolis Rally of 1993. Although the registration number suggested that this was his ex-Monte car, in fact it was a new 'heavyweight' machine, especially built for the occasion.

Boreham had a fine Acropolis record with Escorts, and it repeated the trick in 1993, the cars' performance being helped along by Michelin's amazing new ATS 'mousse' tyres, which took away many of the terrors of high-speed punctures. The result was another triumph – this time for Miki Biasion, in his first World Championship win for Ford, for Michelin, and for the entire Boreham team, which had backed the drivers so well.

Not that Miki was fastest throughout. He carried out a softly-softly campaign, while Sainz (Lancia) and Francois

Ford had not originally planned to contest the Rally of Argentina in 1993, but after a hurried change of plan Miki Biasion was entered and drove a hastily re-furbished test car which Ford had first used in practice for the Monte Carlo Rally. Hampered by the dust clouds generated by preceding cars, Miki took a fine second place.

Delecour shared almost all the fastest stage times. Delecour's car later retired with fuel supply problems, Sainz's machine wilted, Miki's Escort also showed signs of engine troubles, yet he eventually won – by just 73 seconds. It was so typical of the canny Italian – why go faster if you can still win by easing off a little? Miki, accordingly, recorded Ford's third World Championship success in four events.

For Boreham, this changed everything, including the balance of the season. With Ford now matching Toyota, point for point, in the World Championship, extra finance was found to send one car to Argentina, for Miki Biasion to drive. Mounted in a great hurry, this expedition was a success. Fully-loaded Iveco service vans, ex-Acropolis, were shipped out, but the big problem was in finding a rally car for Miki, as Argentina had never been programmed, and nothing was being prepared for it.

In the end, a half-built machine (K57 EVX) which had been used in pre-Monte testing, other testing, and was currently intended to go on a pre-Australia test, was hurriedly completed and air-freighted direct to Buenos Aires. For Biasion, this was normal, as it was the sort of tactic often used by his previous employer, Lancia!

To aid high-speed service operations, Ford Motorsport developed these red quick-lift jack inserts, which plugged directly into the chassis members to help get the front off the ground in a hurry. This is Francois Delecour's car in New Zealand in 1993 on its way to second place overall.

The gamble paid off – almost – as Miki finished close behind the winning Toyota which was running first on the road. He certainly would have won the event if the thick dust which hung interminably in mid-air on many stages had not ruined several of his times, but, as it was, the gap was nearly two minutes, Miki setting seven FTDs but coming second on no fewer than 16 other stages.

Half-way through 1993, therefore, Ford and Toyota

were neck and neck in the Makes Championship, while Miki Biasion actually led the Drivers' table. In truth, this was more than Ford had expected from its brand new car at this point! Then, in the next three events, luck deserted it, and its fortunes slumped.

Two J-registration cars (nominally old, but with brand-new shells) were prepared for entry in New Zealand. They had been finished in May, were shipped half-way round the world, and it was intended that they would be used in Australia, six weeks after New Zealand, then sold off 'down under.' That, however, was Plan A, which rather misfired.

In New Zealand, Biasion crashed his car after completing only three stages, which was very strange, as Miki rarely indulged in off-road accidents, and in spite of setting many fastest times, Delecour could not quite beat Colin McRae's Subaru – he was just 27 agonising seconds off the pace. The

Maybe it was the character of this particular car – J113 BPU – but everyone seemed to get it more sideways than any other works Escort (see page 52). This is Francois Delecour, in New Zealand, 1993.

event was so close, that no fewer than five different cars led at one time or another – and in the end, this was the only World Rally ever to be won by a Subaru Legacy Turbo!

In Australia, therefore, Miki Biasion's bent car from New Zealand had to be rebuilt with the aid of new panels flown out from the UK, but both cars seemed to be mysteriously slow, their engines seemingly down on power and suffering from electronic failures. Neither of them ever figured in the fastest times lists, Miki's rebuilt car was forced out yet again, while Francois's car finished third, nearly four minutes

After it had competed so successfully in Argentina, J113 BPU was shipped part way around the world to compete in the Rally of Australia in September 1993, where Delecour took third place.

Franco Cunico drove this privately-prepared Escort RS Cosworth to an unexpected win in San Remo in October 1993, which proved the versatility of this fine design.

After everything already achieved earlier in the year, another victory in the Rally Catalunya was a real bonus for Ford. Naturally, it was Francois Delecour who helped provide all the excitement.

behind Kankkunen's Toyota and Ari Vatanen's Subaru Legacy Turbo. So from being euphoric after Argentina, Ford then found itself playing catch-up, now many points behind Toyota in the race for the Makes Championship.

Back in Europe, in the San Remo Rally, there was total disaster for the works team but another triumph for the Escort RS Cosworth. Francois Delecour's new car (K201 HNO) stormed off into the lead and then plunged off the road, in thick fog, on Stage 6, while Biasion's engine, in his ex-Acropolis car, burst a water hose and cooked itself just a few minutes later. The Escort RS Cosworth, however, was more successful. Franco Cunico, an ex-works driver from Sierra RS Cosworth days, won the event outright driving a Ford Italy car (the major components of which, including the transmission, had been supplied by Boreham) while Patrick Snyers, in a Belgian-built car, took third place. This meant that the Escort had already notched up four World Championship victories in its first season.

In the meantime, the British works programme of five events had struck problems. Ford contracted Malcolm Wilson's business to build Group A cars for Malcolm and a Group N car for Robbie Head, and SPG also built a Group N car for Gwyndaf Evans. Luck was against Malcolm in 1993, however, for although his car was always fastest, and he usually led events, he only won one rally in K202 HNO, in Ulster. Gwyndaf and Robbie, on the other hand, dominated the Group N category, with Gwyndaf taking the honours at the end of the season. A season which, unfortunately, included a bad crash in the Isle of Man.

For Boreham's Group A cars, fortunes changed for the better in November, when the team tackled two World Championship events in quick succession. Firstly, two tarmac cars, running down to the minimum weight limit allowed by the regulations, started the Catalunya Rally of Spain, in what would decide the Drivers' Championship. Then, one official entry and three works supported machines entered the British RAC event. The character of the two meetings could not have been more different. Catalunya was an all-tarmac affair, in warm Mediterranean weather, where the crews had practised extensively. The RAC was its usual four-day marathon around Forestry Commission territory in England, Wales and Scotland, where reccie crews were merely allowed to drive through the special stages at very low, carefully controlled, speeds.

To keep his Championship hopes alive, Francois Delecour needed to win in Spain, and he set about doing that in his typically flamboyant way. It was a command performance of the type which every Ford enthusiast had come to know, and to admire. Not only did Francois win outright (in L421 NHK), but he set 15 fastest stage times and 11 second fastest, out of 29 stages in all. Total dominance, which no other driver – not even Didier Auriol in the latest Toyota Celica Turbo – could match. It was Francois's third, and the Escort's fifth, World Championship success of 1993. Miki Biaison's car, on the other hand, suffered a turbocharger wastegate failure on one stage, and Miki was never quite as fast as Delecour, so he could only take fourth place.

Then came the RAC Rally, where Delecour's car (another brand-new machine – L422 NHK) was backed by Malcolm Wilson and Robbie Head in two of the British Championship Group A 'Michelin Pilot' Escorts (K202 HNO and K200 FMC), with Gwyndaf Evans's hard-working Group N machine appearing in a Shell colour scheme. Biasion, who did not like the British event at all, was rested, and sent off to France to start testing for the Monte Carlo Rally instead.

On the RAC, which seemed to have more snow, ice and treacherous conditions than ever before, the Escorts performed extremely well. Beaten only by one Toyota and one Mitsubishi, Malcolm Wilson, making up for his unhappy season, finished third overall, beating Francois Delecour in the process and setting three fastest times. Apart from the fact that Robbie Head ended his event off the road, in Keilder forest, (he had been holding down eighth place at the time) it was a good week for Ford, as Gwyndaf Evans captured the Group N category by a huge margin.

To Colin Dobinson, the perfectionist, this had not been a perfect season. Neither his team, nor his star drivers, had lifted the top prizes – they were second in each category – and after seeing the Escort RS Cosworth take five World Championship victories, along with second and third places in abundance, was he content? Certainly not! There were several 'ifs' – but Dobinson did not trade in 'ifs.' If Delecour had won in Monte Carlo (it was only doubtful practice which thwarted him), and if he had not crashed in thick fog in Italy, he could so easily have become World Champion – and Ford could have been the Champion marque once again. So near, and yet ...

Even so, every rabid Escort enthusiast was encouraged, particularly as privately-prepared Escorts (all of them using pieces provided by the Parts Department at Boreham) had also won no fewer than 15 European Championship rounds. For 1994, the cars would assuredly be even better than before, and every Boreham team member now had lots of experience of events worldwide. The assault on the 1994 World Championship promised to be enthralling.

1994

How is it that a team which won the Monte Carlo Rally of January 1994 could be threatened with complete closure in July of the same year, yet bounce back after forging a joint deal with the Belgium rally team RAS? The simple answer is money – or a lack of it. If budgets had not been shuffled

Driving a privately-prepared example, Gwyndaf Evans won the Group N category of the 1993 RAC Rally and finished 11th overall.

around, a great deal of pushing and pulling had not been completed, and a great deal of arm twisting of sponsors and related companies had not been done, Boreham might indeed have closed its doors.

In 1994 there was one big personnel change. Motorsport Director, Colin Dobinson, who had only taken over from Peter Ashcroft at the end of 1991, well before the Escort RS Cosworth was homologated, decided to take early retirement before the end of the season. One story, never convincingly denied by any Ford spokesman, is that Colin was eased out to make way for Peter Gillitzer, a Ford Australia marketing man who had always been vitally interested in motorsport, and had the ear of top management. This was a time when Ford of Europe's CEO was Jac Nasser, who also hailed from Australia – and the business connection was obvious. Gillitzer had already been involved in Ford's F1 supply negotiations, and Nasser then moved him into Boreham.

The new season started well, with Francois Delecour and Miki Biasion confirmed as drivers for the entire season, and, according to the pre-season forecast, with Bruno Thiry of Belgium destined to start seven events (mostly with the support of RAS of Belgium, who had much works support). As in 1993, the works RS Cosworths were strongly sponsored by Mobil. More than ever before, though, the cars also benefited from innumerable minor sponsorship deals, so that the livery and colour scheme on the cars never looked quite as integrated as the stylists would have liked.

Later in the year, unhappily, the team's fortunes sagged considerably. Factors in this must have been the unease caused by several personnel and policy changes, many of which were announced and then forgotten or reversed within months. Although a young Finnish driver, called Tommi Makinen, would win the 1000 Lakes for Ford on his only works drive in an Escort (Tommi would then join Mitsubishi and go on to win not one, but four consecutive World Drivers' Championship titles), Miki Biasion seemed to become progressively disillusioned with the team, and the team with him. Even though he was well into his third season with Ford, the Italian had never integrated in the same way as Francois Delecour, and rarely seemed to show the same spark as when he had won World Championships for Lancia.

Worst of all was that in April, Francois Delecour was badly injured, in a non-rallying crash in a friend's Ferrari F40, while he was out spectating on a minor event in France. With both ankles badly broken, Francois would be out of commission from April to August, and even though he later complained that Ford should have let him return weeks earlier, he never seemed to recover his full performance.

During the season, the team's most senior engineers, John Wheeler and Philip Dunabin, made further improvements to the RS Cosworth's specification. Even wider-track suspension was used on tarmac, and later on gravel-specification cars, anti-lag fuel injection systems were tried out on the exhaust side of the engine, and there was much work on modifications to the integral roll cage layout.

Working alongside Mountune, the team's engine builder, there were continual modifications to the engine, though the compulsory restrictor fitted ahead of the turbocharger inlet meant that no extra peak horsepower, only more peak mid-range torque, could be found. Much experimentation went on in transmission differential settings, while a sequential gearchange system was tried, but later rejected. A six-speed, rather than seven-speed, main gearbox was built and tested, but would not be fully adopted until 1995.

The 1994 season could not have opened in a more successful manner, for there was a magnificent Monte Carlo victory for Francois Delecour's car, which came 41 years after the previous Ford success (Maurice Gatsonides in 1953!) in this high-profile event. Still firmly convinced that he had been robbed of success in 1993 because of the sharp practice of a rival team, Delecour was coldly determined to go one better, and led almost all the way. Even so, the fact that Armin Schwarz (Mitsubishi) set as many fastest stage times, and that Juha Kankkunen's Toyota ran him close, set the tone for the year.

Demonstrably fastest in almost all conditions, Delecour finally made up for the heartbreaks of 1991, when his Sierra

This was the victory that Ford had been waiting for – not only the Escort RS Cosworth's success in the 1994 Monte Carlo Rally, but the first Ford/Monte win since 1953 – more than forty years earlier.

Cosworth 4x4 broke a suspension component on the last stage, and 1993, when Didier Auriol's Toyota achieved a very questionable victory. He set up an emphatic performance, and Miki Biasion backed him up with a fine fourth place.

There was no works entry in Sweden (the FIA's infamous policy of turning long-established events into F2-only rounds was now implemented, and none of the front-line works teams attended), so Boreham's next outing was Portugal, where four works-supported Escort RS Cosworths, and several more competitive but privately-entered examples, took the start. Although two cars were forced out, with engine and engine-related problems, two figured strongly, and Biasion notched up third place, even leading the event for a short time.

No lack of commitment here! Francois Delecour has the Escort's front wheels pointing to his left, but the car is already in the middle of a right-hand hairpin bend!

Did any combination more deservedly win the Monte Carlo Rally of 1994? That was the occasion when Francois Delecour avenged the defeat he suffered in dubious circumstances in 1993, made up for mechanical disaster in a Sierra in 1991, and when Ford could finally erase that 1979 loss when sabotage had foiled it of success, too.

Miki Biasion, twice-World Champion, could count himself unlucky not to have won more events in two years with the works Escort RS Cosworth. Maybe it was because his mercurial partner, Francois Delecour, was so very successful in sister cars. Here is Miki on his way to fourth place in the 1994 Monte.

Sensibly judging that it might not win an event which could be horrifyingly costly to contest, Ford then didn't even enter the Safari. Then, within days came the horrifying news that Francois Delecour had been injured in a road smash between his friend's Ferrari F40 and a Citroën ZX, which was practicing for a local rally. This unfortunate episode would side-line him, throughout the summer and thereafter.

Because no other world-class driver was available to take his place (all the best drivers, of course, were under contract to other teams), a whole series of substitutes, some of whom enjoyed more success than others, had to be drafted in on an event-by-event basis. Those in the second half of the season included Malcolm Wilson, Ari Vatanen, Tommi Makinen, Franco Cunico and Stig Blomqvist. For a team looking to

In 1994, Bruno Thiry's car was works-supported on the Monte Carlo Rally, but sported a very unfamiliar colour scheme. He took sixth place on the event.

Miki Biasion jumping high in L731 ONO on his way to third place in Portugal in 1994.

win the World Championship, this was not ideal – and, in fact, it seemed to be swept off course from that time.

Even Miki Biasion, who became Boreham's de facto team leader from that moment, gradually faded away – in fairness, four engine failures to Escorts in his next five events didn't help.

Right away, of course, Ford missed Delecour's hard-driving flair in the Tour de Corse (which he had won so convincingly the previous year), and the team never looked like winning the event. Fifth and sixth on Corsica's winding tarmac was no sort of reward for the sheer guts which the team showed in such adversity, and even in the Acropolis which followed the cars seemed to be no more competitive, and another 5-6 was the result. Bruno Thiry was faster than the more experienced Biasion in Corsica, but it was soon clear that he was becoming disenchanted with the team and his position in it. His contract, due for renewal at the end of the season, would not be extended.

This was one of the early occasions when a works Escort RS Cosworth appeared in the 'Giesse' colour scheme. Bruno Thiry drove this car into sixth place in Corsica in 1994.

Trying its hardest to restore morale, Boreham then sent three cars all the way to Argentina in June, where Biasion and Thiry were joined by rally-legend, Ari Vatanen. Ari soon showed that he had lost none of his enthusiasm for fast Fords by matching Biasion stage for stage, the two of them quite over-shadowing Bruno Thiry's pace. Driving the same car, though re-furbished, with which Francois Delecour had won the Monte, Ari took a fine third place, Thiry was close behind him in fourth, while the hapless Biasion dropped out, close to the end of the event, with engine failure.

Would it then have made sense to fly the same cars across the Pacific ocean to compete in New Zealand? Maybe it would, but time was certainly tight – so, as a result, two different works Escorts (Miki Biasion in his car

Familiar names, new combination – Ari Vatanen and Fabrizia Pons in this Finnish-registered RS Cosworth, taking fifth place in the Acropolis of 1994.

patently unfit Delecour (who was still limping on barely-healed ankles, spending little time walking when he could sit down!) re-appear in the team, but Vatanen (now 42 years old) was all set to attack Finland as if he had never been away. It was Tommi Makinen's victory, however, on his first and only drive in a works Escort, which inspired the troops. It was not as if his car was anything special – Miki Biasion had already driven it in Corsica and Argentina, and it had made its debut late in 1993 – but Tommi certainly was. Having taken the lead at half distance, Tommi set more fastest special stage times than anyone else, and looked like a World Champion in the making.

Which, of course, he was. Unhappily for Ford, who would dearly liked to have signed him up there and then, Tommi was already ear-marked for fame by Mitsubishi (he would drive a works Lancer on the very next event in San Remo) and, in any case, the works Ford team was still

from Monte Carlo) lined up at the start of the Rally of New Zealand, in Auckland. With Biasion this time paired with Ari Vatanen, great things were expected, but although Vatanen soundly outpaced his team colleague, setting five fastest stage times, neither car made it to the finish. Ari had an accident near the end of the event, when he was extremely well placed, and Miki suffered yet another engine failure. Boreham could only sigh deeply, write it off to Murphy's Law, and look ahead.

All in all, it seemed the team needed a lift, and it duly came in August, in Finland. Not only did a

Ari Vatanen and Fabrizia Pons took third place in Argentina in July 1994, the best works Ford finishers on this event.

Amazing! Tommi Makinen of Finland drove a works Escort RS Cosworth once only – and won the 1994 1000 Lakes Rally. He then went on to win four World Championship titles, all in rival cars.

wavering over its own future, and could not commit any resources at that point.

Although San Remo, which followed in October, would surely have been an ideal event for a fast-recovering Delecour to win, it was not to be. In trouble with the organisers before the start, for so-called illegal practising, he was never as fast as hoped, and he eventually retired, after ten stages, without setting a single fastest time. Miki Biasion, on the other hand, relished the idea of performing in front of his own fans, and against some of his old rivals. Although it was Carlos Sainz's Subaru which lead almost to the finish, and Didier Auriol's Toyota that pipped him at the post, Biasion never gave up the chase. He set several fastest stage times and finally took third place, just 47 seconds behind the winner. The plucky

Miki Biasion on home ground – a dusty stage in the 1994 San Remo Rally, on his way to third place.

Thiry also set two fastest times, was consistent throughout, and finished fourth, just 30 seconds behind Biasion.

Even so, at this point in the rally season, the atmosphere at Boreham was downbeat. After nine events, the team had recorded only two victories, and three third places, but no other minor podium positions – and as a follow-up to the ultra-successful 1993 campaign, this is not what had been expected. One immediate result was that Miki Biasion was told that he was going to be released, which meant that his appearance in the RAC Rally that brought the World Championship year to a close, would be his last in a Ford.

Traditionally, Boreham always entered a veritable fleet of cars in its 'home' event, and on this occasion there were Escort RS Cosworths for Delecour, Biasion, Thiry, Stig Blomqvist, Ari Vatanen and Malcolm Wilson. Even so, it was not a happy event for Ford, as Delecour was excluded for inadvertently cutting a whole section out of an early special stage, Biasion had yet another engine blow up (he looked

Martini sponsored this Escort RS Cosworth, driven by Franco Cunico into sixth place, in San Remo in 1994. Later, of course, Martini would become a major sponsor on the Focus WRC cars.

Maintenance during the RAC Rally of 1994 for Bruno Thiry's car, which would go on to take third place overall.

Having won the 1994 British Rally Championship in this Michelin Pilot-sponsored car, Malcolm Wilson entered it for the RAC Rally but did not reach the finish.

almost relieved when it happened), and even the normally reliable Malcolm Wilson went off the road.

This, in fact, was a period in rallying when Subaru's flat-four Impreza Turbo was maturing, and when Toyota's four-wheel drive Celica seemed to get better and better with every event. The Escort Cosworths, so outstanding in 1993, seemed to be only as fast as their rivals this year, and the RAC result was typical of their season. Thiry, Blomqvist and Vatanen finished 3-4-5, which was at least a solid if not remarkable, result. Even so, in this 'end of season' jamboree, there were other compensations for Ford. Driving a front-wheel drive RS2000 which had been prepared by Gordon Spooner Engineering with a great deal of support from Boreham, Gwyndaf Evans won the new front-wheel drive Formula 2 category (and finished a sparkling seventh

No fewer than six works or supported Escort RS Cosworths started the RAC Rally of 1994. Left to right they were driven by Ari Vatanen, Bruno Thiry, Francois Delecour, Miki Biasion, Stig Blomqvist and Malcolm Wilson.

overall), while Johnny Milner's Shell-sponsored Escort RS Cosworth won the Group N category outright.

1994, too, was a year in which the Escort RS Cosworth won many events, at European and National Championship level. The combination of turbo-horsepower, four-wheel drive, an agile chassis, and the easy availability of competition parts meant that many privately-built cars were out, all round the world. This, indeed, was a season in which Malcolm Wilson's works-specification (but Michelin-backed) Escort RS Cosworths had totally dominated the British Rally Championship – Malcolm won the Keilder-based Pirelli, the Scottish, the Ulster and the Manx outright, while another of his cars won the Rally of Wales – while there were no fewer than 29 Escort RS Cosworth outright victories at European Championship level.

Patrick Snyers, driving a car prepared by RAS Sport, and heavily financed by the Bastos tobacco concern, won the European Championship itself, while other Escorts won the Middle East series, the Mitropa Cup, the North American Zone, and the Austrian, Belgian, Bulgarian, Swiss, German, Estonian, French, Greek, Italian, Dutch, Portuguese, Swedish and Turkish series too.

1995

The results of the 1995 campaign tell their own story. Although Ford finished third in the Makes Championship, the Escorts never won an event, and rarely looked as if they might do so: the team was not the force that it had been in 1993. Although completely recovered from his leg injuries, and with a new co-driver, Catherine Francois, Francois Delecour did not seem to be as determined, or as flamboyant, as he had in the past. With Miki Biasion no longer involved, the team's regular second driver, Bruno Thiry (nominated and backed so faithfully by his Belgian compatriots) was often seen not to be as fast, and certainly not as lucky.

It was the look of the cars, as much as their performance, that told a story too. In 1993 and 1994, the works Escorts had carried flamboyant Mobil livery, and looked purposeful. In 1995, by comparison, they carried a Belgian-inspired

Francois Delecour so nearly added to his 1994 Monte victory in 1995, but in the end had to settle for second place. His team-mate Bruno Thiry (seen here) took fifth.

Francois Delecour finished second in the 1995 Monte – his Escort RS Cosworth rather flamboyantly carrying 'RAS007' stickers, as if the organisers' number plates were not big enough!

Bruno Thiry deserved to win in Corsica, but was cruelly foiled when a wheel bearing collapsed at a place where it could not be reached by Boreham's service mechanics. Francois Delecour – in this car – took second place.

colour scheme, where no fewer than four sponsors all vied for attention. The cars, frankly, looked messy, and you could also say the same about the team's performances, which were neither as smoothly regulated nor as accomplished as in the past.

Throughout the year there was just one stunning performance in the Tour de Corse where Ford should have won, but where the car failed at the last moment. Bruno Thiry led from the very start, setting no fewer than 12 fastest stage times – until suddenly, just two stages from home, a wheel bearing broke up far from service and caused him to retire.

Although Delecour took a storming second place in the Monte, and another fine second place (beaten by just 15 seconds) in Corsica, that was the extent of Ford's success.

The further the season progressed, the less capable the team looked, and by the end of the year, after less than a single season, a divorce was brewing between Boreham and RAS Sport. The performance on the RAC Rally told its own story. Although Bruno Thiry finished fifth, he never set a single fastest stage time and at the end of the event he was beaten in a straight fight, by Alister McRae in a Malcolm Wilson-prepared Escort RS Cosworth with a near-identical specification.

Things began to look more promising in Monte Carlo, where Delecour – driving M511 WJN, a car which would appear several times during the coming season – had the beating of every other competitor except Carlos Sainz, who was in a Subaru Impreza. Delecour led the rally briefly at half distance, set six fastest stage times (only Sainz would

Ford was never fortunate enough to win the Rally of Australia with its Escort RS Cosworths. This was 1995, where Bruno Thiry took sixth place overall. As usual, the Escort's front spoiler took a beating on the rough stages.

beat that) and finished second, backed up by Thiry, in fifth place.

Was this going to be the start of something good? Unhappily, it was not. Delecour did not even finish in Sweden, Thiry could only take sixth place and could set no fastest times, while veteran, Stig Blomqvist, almost had the beating of him in a privately-prepared car. Thiry could only take sixth place in Portugal, while Delecour crashed his works car.

Then came Corsica, which Ford could – and should – have won. Delecour, who had won in an Escort in 1993, went into the event full of confidence. He could not quite match the pace of Thiry, something of a tarmac expert, however, and slightly dropped away when he hit a rock. Thiry, on the other hand, took the lead from the start, set no fewer than 12 fastest stage times (there were only 22 stages) – and was cruelly foiled just two stages from the end, when a front wheel bearing collapsed in a place where restricted-

Tall story department – Francois Delecour (left) and Colin McRae in the middle of the Catalunya Rally of Spain in 1995.

service regulations applied, and he was forced to fall out of the event, clearly distressed beyond all reason. Delecour's second place at the finish was no consolation.

Having trekked all the way round the world to compete in New Zealand in July, only one of the RAS-Ford cars had success. The fact that Delecour had a fever throughout the event did not help matters at all. Two months later, in Australia, the same (re-furbished) ex-New Zealand cars fared no better, with Thiry down in sixth place, and Delecour sidelined after an accident.

Would the Escorts fare better in the Spanish Catalunya Rally, which Delecour had won in 1993? They would not, for against the might of the works Subaru team, Delecour could only take fourth place in the by now well-used M511 WJN, which had started the season.

The final disappointment came on the home event, the RAC Rally, where the RAS-Fords were not only beaten by a phalanx of Subarus, but by a privately-entered Escort RS Cosworth which had been prepared by Malcolm Wilson's own team, and was driven by young Alister McRae. It was a truly downbeat performance, for neither Delecour nor Thiry managed a single fastest stage time, while Malcolm Wilson himself crashed his own self-prepared car.

Out in the wider world, however, it was not all bad news. Once again, at European level, Escort RS Cosworths – some works-prepared, some works-blessed, and all fitted

with a multitude of Boreham components – won no fewer than 27 events outright, which was more than half the total promoted. In addition, although Ford was still not totally committed to developing a full-specification front-wheel drive RS2000, in the RAC Rally itself, Gwyndaf Evans was once again able to win the increasingly important F2 category – being beaten only by five four-wheel drive cars, three Subarus and two Escort RS Cosworths.

(Left) Bruno Thiry in his works Escort RS Cosworth jumping high in the British RAC Rally at the end of 1995; he took fifth place.

Alister McRae, driving one of Malcolm Wilson's privately-prepared Escort RS Cosworths, finished fourth overall – beating the works cars – in the 1995 RAC Rally.

Carlos Sainz's first drive for Ford on his return was in the Swedish Rally where he finished second, just 23 seconds behind the winning car.

1996 – The only way was up

Cometh the hour, cometh the man. At the end of the 1995 season, Ford's rally programme seemed to be in complete disarray. The RAS-Ford link had been dissolved, with very few regrets on either side, and none of the drivers had definitely been re-hired for 1996. With only eight weeks to go to the start of the Monte Carlo Rally, was there any way that Ford could come back?

One man thought there was. John Taylor, the ex-Rallycross Champion, who had been working at Boreham since the end of the 1970s and was now the Operations Manager, had once admitted that 'Ford' would be found emblazoned on his heart when he died. He now moved very swiftly, in the wake of Toyota's technical and managerial chaos, and persuaded one of his all-time heroes, Carlos Sainz, to sign for Ford! John has total recall of this event: "I

There were no works Escorts in the 1996 Monte, so private owner Patrick Bernardini won the event in his own Belgian-registered car.

told Peter Gillitzer that we needed a front-line personality like Carlos to regain credibility. He agreed that I should fly out to Spain, and the rest you know ..."

This dramatic signing came in December, immediately after a series of shaming events saw Toyota Team Europe banned from World Rallying for the whole of the forthcoming (1996) season! As recently as the 1995 Catalunya Rally in Spain, Toyota's cars had been found to be running illegal-specification turbochargers – an efficient, but invisible and quite illegal, method of channelling air around the compulsory turbocharger restrictors, thus increasing the possible horsepower. Before the ban was enacted (and before the scale of Toyota's crimes was known), Carlos Sainz had already agreed to leave his current team (Subaru) and to drive for Toyota in 1996. Accordingly, when the ban was enacted on 3 November 1995, Subaru had re-organised its team for 1996, Sainz found that Toyota could no longer offer him a seat, and he was effectively out of work! John Taylor, who in 1987 and 1988 had managed Sainz's efforts in Sierra RS Cosworths with Ford Spain, swooped at once. He flew out to meet Sainz and spent much time persuading the double World Champion that he ought to sign for Ford.

If Sainz had not actually been on the market with few obvious options, this would not have been easy, for after the

The two works Escort RS Cosworths were specially modified for the 1996 East African Safari. Stig Blomqvist, driving M512 WJN, finished seventh.

1995 season, a seat at Ford – even the team leader's seat – did not look very attractive. Taylor later admitted that he had never worked quite so hard as he did when convincing Carlos, and it worked. Carlos was contracted to Ford in a two-season deal and – more important still – would bring valuable sponsorship with him from the Spanish oil company Repsol.

By any standards, this looked like a new beginning. Once again the works cars would always be based at Boreham, and it would be an Englishman (John Taylor) who ran them on events – but there was still a snag. In recent months, somehow, Ford had fallen out of love with Francois Delecour, whose contract was not renewed. After starting just one event in 1996 in Sweden, he left, and returned to the Peugeot camp.

Accordingly, until Bruno Thiry once again became a regular team member (from June 1996) there was a constant and somehow unsettling turnover of guest and one-off drivers, though those 'guests' – Stig Blomqvist and Ari Vatanen in particular – were distinguished by any standards.

Ford's new season started with an enormous surprise. Because the Monte Carlo Rally was not a World Championship qualifier in 1996 (the FIA's policy of 'rotating' qualification – too complex to explain here – was in full swing), there were no factory Fords in the Alps in January. No matter. Alliance Yacco Ford entered an Escort RS Cosworth, RAS Sport built it, and Patrick Bernadini drove it – the result was a comfortable and emphatic outright victory!

Sainz, in the meantime, had started in the way he meant to go on. Within 24 hours of signing his new contract with Ford, he was out testing in Sweden, and for the rest of the season he would bring his enthusiasm and determination to bear on a team that needed every scrap of encouragement to get back on track.

It worked. Although Carlos would win only one event during the season – that being the Rally of Indonesia – he would also take three second places and two thirds. The car would only let him down twice – with a suspension failure in the Safari, and a transmission breakdown in the 1000 Lakes – and his bright, Repsol-liveried Escorts always

Although he was used to winning World rallies, there was real joy for Carlos Sainz in winning the 1996 Rally of Indonesia in this Escort RS Cosworth.

Though not the fastest on most stages, Carlos Sainz was as consistent as ever, and took victory in the Rally of Indonesia in 1996, just 23 seconds ahead of his nearest rival.

looked likely to win. At the end of the year, Carlos took third in the Drivers' Championship (with one fewer breakdown he would certainly have been second) and the image of 'his' team had been totally transformed.

And it was his team, for sure. Never before had Ford relied so much on one personality to give them hope – and Carlos certainly delivered. The cars were no faster than before, but they were more reliable – the difference was in the team, and how it performed. Although there were no major breakthroughs, Philip Dunabin's group of designers

Jarmo Kytolehto's Escort RS Cosworth took a sensational third place in Finland in 1996, the best private entry result in full World Rallying for several years. The car had been prepared by Malcolm Wilson Motorsport.

Martin Whitaker directed Ford Motorsport's fortunes from mid-1996 until the early 2000s, this period including the two seasons in which the Escort WRC was used.

refined almost every aspect of the layout. More responsive engines, active differentials at the front and in the centre, stronger suspension components, new-type Dynamic shock absorbers and re-located fuel tanks all featured during the season – even though this was effectively the 'Indian summer' of the Escort RS Cosworth. Dunabin could now admit that the centre of gravity was too high, that the engine was too heavy (Ford was the only team still using iron-block engines), and that the rear aerodynamic spoiler wasn't very effective – but he also knew that improvement, and a new model, the Escort WRC, was being planned.

For Sainz, in what was now seen as an ageing design, the whole season was an uphill battle against Mitsubishi and Tommi Makinen, but he never gave up. Second to Tommi in Sweden by only 23 seconds, he then won in Indonesia in spite of suffering two crashes and a broken gearbox. Third in Greece and second in Argentina – both rough and tough events – showed that the Escort might be a touch heavy, but it was still very strong, while third in Australia was achieved in spite of the car drowning out for some time, in a deep watersplash mid-stage!

By seconds, and the odd place, though, it was not quite Carlos's year. Although he set more fastest stage times than Colin McRae on the San Remo, he was still defeated by just 22 seconds. This had happened so often during the season that Ford realised that the Escort RS Cosworth's glory days were over. Morale, however, was strong. Results were once again improving, and the team knew that success was due in 1997.

What goes up must come down – Carlos Sainz in Australia in September 1996. The result? Third place, the best of the works Escorts in this rally.

San Remo in 1996, with Carlos Sainz so very close to beating Colin McRae's car for outright victory. In the end, Carlos took second, just 22 seconds behind Colin.

In Britain, and in Europe, too, there was great joy. In the British Championship, Gwyndaf Evans's F2 Escort RS2000s (cars which ran with a great deal of factory support and technical expertise) won the Series at a canter. In the European Rally Championship, no fewer than 24 of the 53 qualifying events – nearly half of them – were won by privately-entered Escort RS Cosworths, running with Boreham-developed pieces.

Martin Whitaker (right) took over as Motorsport Director from Peter Gillitzer in late 1996, and immediately authorised work on the Escort World Rally Car. Here he is, with Cosworth's then Managing Director, Dick Scammell.

1997

By mid-1996, the wind of corporate change was roaring through Boreham. Not only was a new version of the Escort being developed (though this would not affect Ford Motorsport's rally programme for two more years) but there was to be a new Director of Motorsport – and there were renewed rumours about Boreham's actual rally team, as opposed to its development facility, being completely closed down.

At the end of his two-year secondment to Boreham, Peter Gillitzer's contract was not renewed, and his place was taken by Martin Whitaker. The change of atmosphere was immediate. Not only was Whitaker the archetypal, smooth speaking, public school-educated British manager whose CV included previous high profile appointments at the RAC MSA, at the McLaren F1 team, and at the FIA, but he came to Motorsport from Ford's Public Affairs division, and was already well-respected in the 'corridors of power' at head office!

Amazingly, he also seemed to know where the money was buried – money that had not previously been made available to Ford Motorsport. The prospects for a resurgence at Boreham were bright, and there was more. Project work had already started on a car we later came to know as the Escort World Rally Car – and if John Taylor got his way, there would be additions to the technical and driving strength too!

This was a was a season in which Ford came so close to the ultimate glory – where the team finished second in the Makes Championship, and, as everyone expected, the consistent and charismatic Carlos Sainz took third place in the Drivers' series. Carlos himself won twice – in the Acropolis and in Indonesia – and backed this up with second places in Monte Carlo, Sweden, the Tour de Corse and New Zealand. Not only that, but in Corsica he was foiled by a mere eight seconds, in New Zealand by 13 seconds, and in Sweden by 16 seconds. His WRCs broke under him in the Safari, Portugal, Argentina, Finland and Australia – no wonder that Carlos knew that he had been capable of winning the Championship.

Alongside Carlos, Armin Schwarz of Germany started the first six events of the year, and notched up a third and two fourth places, though he was never to be as rapid as Sainz. When it became clear that sponsorship funds that he had promised as part of the deal for 1997 would not arrive, Ford speedily released him from a contract which it saw as null and void.

It was quite by chance that his replacement, four-times World Champion, Juha Kankkunen, was on the market without a drive, for he should have been occupying a Toyota seat, yet that company was still not allowed return to World Rallying after being suspended for homologation infringements in the immediate past.

Juha therefore joined Ford Motorsport in May (it was the first time he had ever had any links with Ford), soon came to terms with the vice-free handling of the Escort WRC, and was immediately on the pace. By the end of the season with Ford, he had recorded no fewer than four second places. On two of those rallies – both of them on loose-surfaced events – he finished very close behind team-mate Sainz (by 7 seconds and 16 seconds, respectively) and might easily have won, but as Ford was trying to support Sainz towards the World Championship, team orders were probably applied.

All in all, works cars, as entered by M-Sport, started all 14 World Championship rallies, and scored in 11 of them. The biggest miracle, however, was undoubtedly that M-Sport actually built two newly-designed cars in less than two months before the Monte Carlo Rally, made them reliable, and started producing results almost at once.

As expected, Sainz did his professional best right from the start, and in spite of suffering an engine which misfired from time to time, and setting only one fastest stage time, he was 'Mr Consistency', and finished second to a works Subaru Impreza WRC. Armin Schwarz, however, crashed his car, and twisted the chassis – which didn't please M-Sport, who, as yet, had no spare cars with which to carry on rallying – though he did go on to take fourth place, and even

Juha Kankkunen joined the works team in mid-1997, and became team leader in 1998.

set two fastest times! It was the same story, different only in details, in Sweden, where Carlos yet again took second – but on this occasion, he set four fastest times, and was only defeated by 16 seconds.

Against all the odds – and all the forecasts – M-Sport had two Escort WRCs ready to tackle the 1997 Monte Carlo Rally, where they finished second and fourth. This is Carlos Sainz on his way to second place.

For Ford, and M-Sport, this was almost a fairy-tale beginning, which not even a Hollywood scriptwriter might have envisaged, especially as it was Schwarz's turn to finish well up (fourth, but a long way off the pace) in the Safari. Sainz's car, unhappily, was forced to retire when a wheel came off, after several wheel studs had sheared, and he was too far from any feasible service car to be able to drag his car along on three wheels.

Jumping high in Sweden on roads which had surprisingly little snow, Carlos Sainz's WRC was on its way to another second place just 16 seconds behind the rally winner.

Still using its early-season Monte and Sweden cars, the works team travelled to Portugal, where Schwarz finished third and Sainz retired, this time with a seized gearbox. Three weeks later and over the border in Spain, the team hoped for better luck, especially as team-leader Carlos was usually treated like royalty by his many fans. After M-Sport made engine specification changes in testing, however, both power units suffered serious overheating, both had burnt-out exhaust pipes, both overcooked the transmission joints close to them, and both suffered accordingly. No success in Spain, therefore.

Corsica was better, much better, for Carlos Sainz, who used his repaired Catalunya car to great effect. The exhaust systems had been revised and the engines had received

Ford built up two special 'heavyweight' WRCs for the 1997 Safari, and it was Armin Schwarz in this car who finished fourth.

Once again, Carlos Sainz had to settle for second place in a World Championship Rally, this time in Corsica in 1997 where he missed out by a mere eight seconds!

much attention, and the combination was more effective than expected. This was one of those 'if only' occasions, where one bit of trouble for the main competition (Colin McRae's Subaru) would have made all the difference. Carlos led the rally on several occasions, including going into the final stage where everything revolved around tyre choice, which McRae got exactly right. McRae won that stage by 15 seconds, and the rally itself by just eight seconds! Carlos was downcast, and rightly so.

So was Armin Schwarz, who was dismissed immediately afterwards because the sponsorship funds he had promised when being hired had still not turned up, and M-Sport was extraordinarily lucky to find Juha Kankkunen available. The ex-World Champion signed up, joined the team for the next

Team work! Carlos Sainz (right) and his co-driver, Louis Moya, celebrate victory in the Acropolis Rally of 1997, with M-Sport team boss, Malcolm Wilson.

event (Argentina) and was immediately seen to be Carlos's equal on loose surfaces. Very encouraging.

Not that this helped in Argentina, where both cars repeatedly damaged their vast front bumper mouldings when ploughing through deep water. (M-Sport went to the trouble of having replacement moulding fabricated overnight by a local motor racing specialist company so that it stayed within the regulations.) It was only coincidence that both cars subsequently retired, one with damaged front suspension, the other with engine failure.

Then came Greece, and the Acropolis, where Ford's traditional strength (literally) in this tough event, was confirmed yet again. Once the challenge from the Subarus and Mitsubishis had wilted, the Escort WRCs were comfortable leaders, with Sainz taking no fewer than 12 fastest stage times, closely tracked by Kankkunen in almost every case. Left to their own devices, the two Ford drivers could have fought to the end – and Kankkunen might have won – but M-Sport applied team orders. The Finn loyally tucked in behind and the result was that Carlos won the event by 17 seconds from him, but by five minutes

Where else would you find such a high jump other than Finland in the Neste/1000 Lakes Rally of 1997 with Juha Kankkunen's Escort WRC on its way to second place, just seven seconds behind the winning car.

from Tommi Makinen's Mitsubishi. It was the Escort's first 1-2 result since 1993.

Meanwhile, the Argentina cars had been shipped directly out to New Zealand where they were re-prepared, so well, in fact, that Sainz once again took second place (he was getting used to this) losing out by a mere 13 seconds, with Kankkunen right behind. Carlos would certainly have won if he hadn't unfortunately struck a sheep on the last day, badly damaging the nose of the Escort, and slaughtering the hapless animal. Then, in Finland, it was Juha's turn to finish second, just seven frustrating seconds behind Tommi Makinen's winning Mitsubishi, though Sainz's car wilted with transmission failure.

Although this was indeed an intensive season, M-Sport

then sent two already hard-used cars (P8 FMC and P9 FMC) to Indonesia, where it was overjoyed to register a second 1-2 finish. Although McRae's Subaru led the event up to halfway, Colin then had a big accident which damaged the engine cooling system, and that was that. The Escorts took turns to lead, Sainz got the final nod from Malcolm Wilson, and won by just 14 seconds.

That, though, was the high point of the season, for they were out-paced in San Remo (Sainz, fourth, and Kankkunen, sixth) and the ex-Indonesia cars were both shipped again,

The 1997 Rally of New Zealand was incredibly hard-fought. Juha Kankkunen, driving this Escort WRC, took third place, just six seconds behind team-mate and rival Carlos Sainz who himself was only 13 seconds off an outright victory.

This works service point was busy during the 1997 Rally of New Zealand, with Carlos Sainz's car (foreground) and Juha Kankkunen's car both in for work. Compared with F1 stops, a rally car halt is rather messy – but just as much work gets done!

Near-jungle conditions in Indonesia, in 1997, with Carlos Sainz on his way to a well-deserved victory. Team-mate Juha Kankkunen would take second place.

When Carlos Sainz and Louis Moya won the Rally Indonesia in 1997, it was a repeat performance for they had won the same rally a year earlier. The difference this time was that they were using an Escort WRC.

re-prepared, and then retired in Australia. Carlos's car suffered engine failure, after he had taken seven fastest stage times, while Kankkunen went off the road and rather severely hit a tree.

This left Ford with just one event – the British RAC – to round off the year. Because both the Manufacturers' and the Drivers' titles had already been settled, it was only pride which was at stake, and on a compact 26 stage event, M-Sport was hoping for victory. In the end it was close, but not quite close enough, for although the two newest Escort WRCs (R1 FMC and R2 FMC, which had made their debuts in Finland) appeared, neither was quite a match for Colin McRae's Subaru Impreza WRC. As it was, they took four fastest stage times between them, and finished solidly second (Kankkunen) and third (Sainz) overall.

For M-Sport, it had been a remarkable first season, if not quite rewarded as the team would have wished. The records showed that no fewer than 27 new Escort WRCs had been built during the year, 11 of them at M-Sport in Cumbria, almost all for its own use, and eight more at Boreham, for sale to private customers.

Other privately-owned WRCs were soon completed, and started

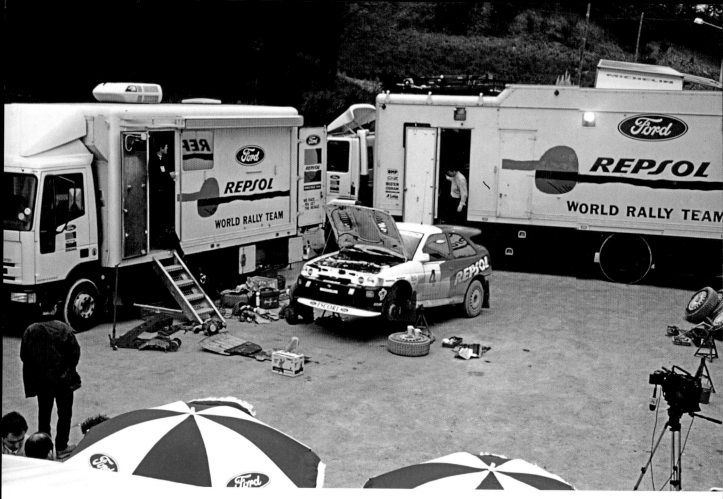

Although Ford's service 'umbrella' was perhaps not as extensive as some of its Japanese rivals, the drivers lacked for nothing. This was 1996, a year in which Carlos Sainz's cars always carried the competition number four.

rallying, not only in World, but in European events in various countries. Patrick Snyers's Bastos-backed Belgian car was extremely successful – it won European Championship rounds in Belgium and in Poland, while Ari Vatanen drove it in the RAC Rally. There was also a magnificent performance in the Middle East Championship, where Mohammed Ben Sulayem dominated the series in Escort RS Cosworths, and later in Escort WRCs.

Not only that, but private entrants in the older Escort RS Cosworth, which now counted as an obsolete design, notched up a further six European Championship outright victories

All in all, the Escort WRC, which was admittedly only a stop-gap design, and likely to be the very last of this long-running rally car series, had had an amazingly successful first year. Without spending anything like the same amount of money as their heavily-backed Japanese rivals, Ford and M-Sport had delivered on every promise.

A both proud and sad occasion, for this would be the last time that 'King Carlos' (Carlos Sainz) – would ever drive an Escort WRC in motorsport. In the 1997 Network Q Rally of Great Britain, Carlos took third place in R1 FMC

The M-Sport Escort WRCs had a new livery for 1998, this being Juha Kankkunen on his way to second place in the Monte Carlo Rally.

1998

For 1998 the entire world of rallying, it seemed, knew that the Escort WRC would be starting its second, and last, World Championship campaign. Rumours, planted stories and sneak pictures of a new-generation Escort replacement had already been found in the motoring press, and M-Sport was preparing for its own future. Months before the road car was even previewed, M-Sport had started work on the design of its Focus WRC, a car which would share nothing with the Escort, and which would be engineered in secrecy at the Millbrook 'proving grounds,' in Bedfordshire. Once this task began, there was little time, and no funds, available for the further development on the Escort WRC.

M-Sport started the new season with a different

turbocharger housing, which helped to provide a more powerful engine, and the cars became so light that they needed ballast to run at the minimum weight being imposed by WRC regulations. Electronic 'launch control' of the engines was used to give even more effective starts from the beginning of special stages.

There were important driver changes too. Everyone at Ford was sorry to see Carlos Sainz leave the team at the end of 1997 (he would be back – in the Focus WRC – in 2000), for he had gone off to Germany to renew his long-time, on-off-on love affair with Toyota. I should record that he would come close to winning the 1998 Drivers' Championship with his new team.

This meant that the Finn, Juha Kankkunen, became Ford's team leader. Juha was still formidably fast, but no longer quite as driven, or quite as dedicated to detail improvement, as Carlos Sainz had been. Juha was also known to be less comfortable tarmac rallying than he was

The usual dust clouds followed rally cars wherever they went in Argentina, in 1998: Juha Kankkunen was on his way to third place in R1 FMC.

on the loose. For its second driver, Ford reverted to old times, bringing back Bruno Thiry who had driven Escort RS Cosworths on so many occasions in the mid-1990s.

For 1998 there was also a change of sponsorship, and therefore of livery. In 1996 and 1997, the major Repsol sponsorship had come from Spain, and was quite naturally linked to Carlos Sainz. It was logical that when Carlos moved on, the patronage should therefore follow him to his new team (Toyota). Accordingly, for 1998, the works Fords looked very different – for they were now to be blue and white, with the corporate Ford 'Blue Oval' much in evidence, along with new oil company support – from Valvoline.

As one might have expected, this was to be a relatively quiet year for Ford. Once the specification was settled at the start of the year, there were virtually no technical novelties throughout the season. Unhappily, too, the 1998 model was no great advance over that of 1997, so it was less competitive against the new competition.

In its final full works season, the Escorts would finish fourth in the World Rally Championship, behind three well-funded Japanese rivals, all with modern machinery. Although there were no outright victories to celebrate this swan song, there were any number of fine performances, and podium positions.

Juha Kankkunen took fourth in the Drivers' Championship, and was always extremely consistent. He took three second places (in the year's most high profile events – Monte Carlo, Safari , and the RAC Rally of Great Britain) and was third four times more – as expected, every single one of those fine performances was a loose surface, or a winter, rally.

Bruno Thiry was neither as fast, nor as explosively successful, as his Finnish team-mate. He recorded just one third place – on the RAC Rally of Great Britain – but was unlucky, suffering from four mechanical breakdowns, three of them engine-related.

One happy occasion which kept Escort nostalgia well and truly alive was when Ari Vatanen was invited to compete in the Safari, as a stand-in, after Bruno Thiry broke his ribs in a bizarre practice accident when he was actually

Juha Kankkunen, high off the ground, on one of Finland's famous jumps during the Neste Rally. By this time, the Escort WRC was at the limit of its development, though still recorded third place.

travelling in the rear of another training car! It was twenty-one years since Ari had first tackled the Safari, and ten years since his previous entry there. Taking to the Escort WRC as if he had only been away from Ford for a few days (Ari had driven Escort RS Cosworths in the mid-1990s) he seemed to be equally as fast as Kankkunen throughout, and eventually finished third, just 25 seconds behind his fellow Finn. Ari also took fifth in Portugal (an event where Kankkunen should have been second, before a transmission problem dropped him way down), before handing back his seat to Bruno, whose ribs had mended.

It was fitting and nostalgic but, above all, sad, that the works Escort team should start its last front-line team event on home ground. The headline writers, however, could not have a field day, for this was not quite like old times. In a consistent, though not super-fast, display, Juha and Bruno finished second and third overall in the Rally of Great Britain. None of the less hard-bitten reporters missed

(Above) Water, water everywhere – so typical of conditions on the East African Safari of 1998, with Ari Vatanen on his way to a fine fourth place.

On the Acropolis Rally of 1998, Juha Kankkunen took a fine third place, as in 1997, just an agonising 44 seconds away from outright victory.

Bruno Thiry hurled his works Escort WRC to the very limits of adhesion on the Rally of Great Britain in 1998. Bruno took third place and his team-mate, Juha Kankkunen, recorded second in the WRC's last serious appearance in World Rallying.

This was positively the very last works appearance by an Escort WRC – with Petter Solberg and Phil Mills on their way to a promising 11th place in the Swedish Rally of February 1999. Escorts of one type or another had been rallying for 31 years.

this, for it was the very best which could be expected of this ageing design, especially one which had received no on-going development during its last season. I should also note that Sebastian Lindholm (in a privately-sponsored, but newly-built Escort WRC), took fifth place, while Armin Schwarz finished seventh in a 1997 car which had been prepared by RED. It was fitting, no doubt, that on this event, no other marque was so well placed.

For the record, no fewer than 19 Escort WRCs appeared in the 1998 World Championship season, of which only four had previously featured in 1997. Of these 19, only nine different cars were official M-Sport entries. When a number of private conversions, which appeared in lesser events throughout the season, are added in to the total, this brings the total of Escort WRCs built in two years to well over 50. Not bad for a stop-gap design ....

At European Championship level, there were six outright Escort WRC victories, some of them by one-time works driver, Franco Cunico, based in Italy and using a Martini-sponsored machine. For Ford, success like this was commercially vital, as Martini (the Italian drinks giant) was about to join forces with M-Sport, for the Focus WRC World Rally programme of 1999. Escort RS Cosworths won seven times in European events too, and once again, Ben Sulayem won the Middle Eastern series very easily indeed.

There was one very important social occasion too. Immediately before the Rally of Great Britain, which brought the rallying year to a close, Ford chose to celebrate the imminent retirement of the works Escorts, by hosting a glittering celebratory dinner for every important Ford personality who could be persuaded to attend.

In an occasion which has never been matched by any other team, Ford attracted no fewer than six World Rally Champions to attend – Bjorn Waldegård, Ari Vatanen, Hannu Mikkola, Juha Kankkunen, Carlos Sainz and Tommi Makinen. Timo Makinen (a triple Escort rally winner), Ove Andersson, David Richards, Malcolm Wilson and Andrew Cowan were also there. Walter Hayes, Stuart Turner, Martin Whitaker, Bill Barnett and many others joined them – and there was regret that Peter Ashcroft (who, by this time, had

What a great occasion! When Ford officially decided to retire the Escort from rallying, replacing the car with the new-generation Focus, it threw a magnificent party at Sudeley Castle, near Cheltenham, to celebrate three decades of success. Apart from World Rally Champions Juha Kankkunen, Tommi Makinen, Hannu Mikkola, Carlos Sainz, Ari Vatanen and Bjorn Waldegård, and team bosses, Walter Hayes Stuart Turner, and Martin Whitaker, how many more famous faces can you spot?

retired to live in the United States) could not be there. It was an honour for all of us to be present, yet there was even more special pleasure in receiving a framed picture of a group of personalities, all of whom had signed their names around the margins. An unforgettable evening.

1999

Even though the Focus WRC had been homologated on 1 January 1999, and Ford's motorsport future was tied up in that car, the career of the Escort WRC was not yet over. In Sweden, in February 1999, the works team's new 'apprentice', Petter Solberg, finished 11th overall in an M-Sport Escort WRC (R6 FMC – Kankkunen's Rally of Great Britain car). But that really was all, and no further works Escort outings took place. Only two weeks after the Swedish Rally, Colin McRae took the Focus WRC to its first-ever victory, in the Safari.

It was in Sweden, therefore, in February 1999, that the rallying career of the works Escorts was brought proudly to a close, almost 31 years after it had begun in San Remo, Italy, in March 1968. This was a period in which countless – tens of thousands, certainly – rallies, rally-sprints and rallycross events had been won by one or other of the Escort family. It was a breed of car which had brought success, sparkle, excitement and glamour to motorsport all over the world – and, in my opinion, no other rally car has ever been as successful. I doubt if the Escort's record will ever be beaten.

World Rally success

From 1993 to 1996, all cars were Escort RS Cosworths. In 1997 and 1998, all cars were Escort World Rally Cars.

Event	Position	Drivers
1993		
Monte Carlo	2nd	F Delecour/D Grataloup
Monte Carlo	3rd	M Biasion/T Siviero
Portugal	1st	F Delecour/D Grataloup
Portugal	2nd	M Biasion/T Siviero
Tour de Corse	1st	F Delecour/D Grataloup
Acropolis	1st	M Biasion/T Siviero
Argentina	2nd	M Biasion/T Siviero
New Zealand	2nd	F Delecour/D Grataloup
Australia	3rd	F Delecour/D Grataloup
Spain (Catalunya)	1st	F Delecour/D Grataloup
RAC	3rd	M Wilson/B Thomas
1994		
Monte Carlo	1st	F Delecour/D Grataloup
Portugal	3rd	M Biasion/T Siviero
Argentina	3rd	A Vatanen/Ms Fabrizia Pons
1000 Lakes	1st	T Makinen/S Harjanne
San Remo	3rd	M Biasion/T Siviero
RAC	3rd	B Thiry/S Prevot
1995		
Monte Carlo	2nd	F Delecour/Ms CFrancois
Tour de Corse	2nd	F Delecour/Ms C Francois
1996		
Sweden	2nd	C Sainz/L Moya
Indonesia	1st	C Sainz/L Moya
Acropolis	3rd	C Sainz/L Moya
Argentina	2nd	C Sainz/L Moya

Event	Position	Drivers
Australia	3rd	C Sainz/L Moya
San Remo	2nd	C Sainz/L Moya
San Remo	3rd	B Thiry/S Prevot
Spain (Catalunya)	3rd	B Thiry/S Prevot
1997		
Monte Carlo	2nd	C Sainz/L Moya
Sweden	2nd	C Sainz/L Moya
Portugal	3rd	A Schwarz/D Giraudet
Corsica	2nd	C Sainz/L Moya
Acropolis	1st	C Sainz/L Moya
Acropolis	2nd	J Kankkunen/J Repo
New Zealand	2nd	C Sainz/L Moya
New Zealand	3rd	J Kankkunen/J Repo
1000 Lakes	2nd	J Kankkunen/J Repo
Indonesia	1st	C Sainz/L Moya
Indonesia	2nd	J Kankkunen/J Repo
RAC	2nd	J Kankkunen/J Repo
RAC	3rd	C Sainz/L Moya
1998		
Monte Carlo	2nd	J Kankkunen/J Repo
Sweden	3rd	J Kankkunen/J Repo
Safari	2nd	J Kankkunen/J Repo
Safari	3rd	A Vatanen/F Gallagher
Argentina	3rd	J Kankkunen/J Repo
Acropolis	3rd	J Kankkunen/J Repo
1000 Lakes	3rd	J Kankkunen/J Repo
RAC (GB)	2nd	J Kankkunen/J Repo
RAC (GB)	3rd	B Thiry/S Prevot

Works rally cars (and when first used)

These are the nominal identities of the factory-prepared (not merely factory-supported) Escort RS Cosworth/Escort World Rally cars, built and registered between 1993 and 1998, for use in major World events. On Escort RS Cosworths (but not WRCs, the regulations were tighter), as bodyshells wore out, or after a serious crash, some of their identities were 'cloned' onto newly-built cars. In many cases, therefore, the use of a particular registration number was often no more than theoretical! A typical example occurred early on – K746 GOO and K748 GOO started life as Monte Carlo tarmac cars in January 1993, but were re-born as Acropolis heavyweights in May, three months later.

On Escort World Rally Cars, new WRC regulations ensured that registration numbers could only be used with the original structure. Specifically, and according to the regulations, if the roll cage survived, so did the identity of the car itself. Even so, I have noted important victories with the identity carried by a particular car.

Escort RS Cosworths

1993
K746 GOO (1993 Acropolis)
K748 GOO (1993 Portugal)
K831 HHJ
K832 HHJ (1993 Tour de Corse)
K57 LVX
L421 NHK (1993 Catalunya/Spain)
L422 NHK

1994
L730 ONO (1994 Monte Carlo)
L731 ONO
L732 ONO
L985 OJN
L124 PAR (1994 1000 Lakes)
L125 PAR
M295 RHH

1995
J204 GPN

M511 WJN
M512 MJN
M513 MJN
M743 YWC
M245 AHK
N788 WHH

1996
N506 EAR
BK872RI (registered with local
    Indonesian identity) (1996
    Indonesia)
BK868RI (registered with local
    Indonesian identity)
B2110MV (registered with local
    Indonesian identity)
M40 FMC
N704 FAR
N705 FAR
N5 FMC
N6 FMC

Escort World Rally Car

1997
M10 MWM
P6 FMC
P7 FMC
P8 FMC (1997 Acropolis)
P9 FMC (1997 Indonesia)
P11 FMC
R963 DHK
R1 FMC
R2 FMC

1998
P221 HAO
R3 FMC
R4 FMC
R5 FMC
R6 FMC
S13 FMC

Index

Note: There are so many mentions of the Escort RS Cosworth and the Escort WRC which took over from it, along with the Motorsport centre at Boreham, that it is impractical to index them.

Amblard, Marc 30
Andy Rouse Engineering 41
Ashcroft, Peter 9, 12, 32, 36, 41, 43, 44, 47, 55, 75, 118
Aston Martin (and models) 27, 41, 43, 45
Audi (and models) 13, 46
Auriol, Didier 17, 57, 60, 64, 73, 76, 82
Austin-Rover (and models) 46
Autosport 45, 57

Barnett, Bill 32, 118
BMC (and models) 42, 43
British Rally Championship 42, 46, 87, 99
British Touring Car Championship 41

Carlsson, Erik 43
Citroën (and models) 78

Dobinson, Colin 36, 40, 44, 46, 54, 56, 73, 75
Dunabin, Philip 27, 29, 36-38, 41, 54, 75, 96, 97

European Rally Championship 48, 52, 87, 90, 99, 112, 118

Ferrari F40 75, 78
FIA/FISA 9, 14, 21, 28, 29, 38, 54, 65, 76, 95

Ford models:
  ACE/CE14 (see Escort RS Cosworth) 11, 12, 14, 19, 21, 23, 26, 43, 45
  Capri RS2600 43
  Escort Mk I and Mk II 11, 12, 41-43, 46, 48, 52, 62
  Escort Cabriolet 21
  Escort RS1600i 43, 46
  Escort RS1700T 43, 45
  Escort RS2000 40, 48, 49, 86
  Escort RS Turbo 23, 25, 43
  Fiesta 22
  Focus (and Focus WRC) 28, 38, 46, 47, 49, 84, 114, 115, 119
  Merkur (Sierra) XR4Ti 21
  Mondeo 22, 30, 41
  RS200 9, 12-14, 43, 45, 46
  Sierra RS/RS500 Cosworth 9, 22, 43, 45, 60
  Sierra Cosworth 4x4 9, 10, 12, 14-16, 22-25, 35, 36, 38, 40, 44, 45-48, 51-53, 57, 73, 75, 77
  Sierra XR4x4 14, 22
Ford premises and workshops:
  AVO 43
  Boreham 9, 11, 15, 21, 27, 29, 30, 36, 38, 40-48, 54-57, 59, 60, 62, 65, 66, 67, 73-76, 79-81, 84, 86, 88, 91, 92, 95, 99, 100
  Cologne 43
  Dunton 23, 36, 45
  Karmann (Germany) 16, 21, 34, 43
  Merkenich 30
  Millbrook 114

M-Sport 29-31, 38, 40, 42, 46, 101-103, 105-107, 112, 114, 118, 119
Ford race and rally drivers:
Aitken-Walker, Louise 46
Andersson, Ove 32, 118
Bardolet, Mia 25, 34, 51, 53
Ben Sulayem, Mohammed 112
Bernardini, Patrick 93, 95
Biasion, Miki 40, 44, 47, 48, 52, 57, 58, 60, 62, 64, 66-70, 72, 73, 75-77, 79-84, 86
Blomqvist, Stig 78, 84, 86, 89, 94, 95
Cunico, Franco 71, 73, 78, 84, 118
Cowan, Andrew 32, 118
Delecour, Francois 40, 44, 47, 48, 51, 52, 57-70, 72, 73, 75-81, 84, 86-90, 95
Droogmans, Robert 41
Evans, Gwyndaf 54, 73, 74, 86, 91, 99
Francois, Catherine 87
Head, Robbie 54, 73
Grataloup, Daniel 61, 63
Kankkunen, Juha 17, 30, 32, 48, 49, 72, 75, 101, 105-111, 114-119
Kytolehto, Jarmo 97
Lindholm, Sebastian 118
Lovell, Mark 46
Makinen, Timo 118
Makinen, Tommi 32, 48, 75, 78, 81, 82, 97, 107, 118, 119
McRae, Alister 88, 90, 91
Mills, Phil 118
Mikkola, Hannu 11, 13, 32, 118, 119
Moya, Louis 106, 111
Milner, Johnny 87
Pons, Fabrizia 81
Richards, David 32, 118
Sainz, Carlos 9, 17, 30, 32, 41, 42, 46, 48, 49, 58, 66, 82, 88, 92, 93, 95-111, 113, 115, 116, 118, 119
Schwarz, Armin 30, 75, 101, 102, 104, 105, 118
Siviero, Tiziano 47

Snyers, Patrick 73, 87, 112
Solberg, Petter 33, 118, 119
Thiry, Bruno 41, 49, 75, 78-80, 84, 86-91, 95, 116, 118
Vatanen, Ari 32, 45, 48, 62, 72, 78, 80, 81, 84-86, 95, 112, 116-119
Waldegård, Bjorn 32, 118, 119

Gatsonides, Maurice 75
Geneva Motor Show 21
Gilbey Engineering 43
Gillitzer, Peter 36, 40, 41, 46, 75, 93, 100
Gordon Spooner Engineering 40, 51, 86
Griffiths, John 9, 54, 55

Hart, Brian 43
Hayes, Walter 10, 32, 42, 43, 118, 119
Haynes of Maidstone 45
Hodgson, Melvin 54
Howe, Bob 43

Jaguar (and models) 43
Jones, Mick 55

Lancia (and models) 9, 11, 14, 16, 17, 21, 22, 41, 47-49, 58, 66, 67, 75

Makinen, Tommi 17, 32
Malcolm Wilson Motorsport 97
McRae, Colin 17, 49, 60, 69, 90, 98, 99, 105, 108, 111, 119
Meade, Bill 9
Mike Taylor Developments 34, 50
Mitsubishi (and models) 16, 17, 48, 57, 73, 75, 81, 106, 107
Moreton, Mike 9, 19, 21, 43
Motoring News 42
Mouton, Michele 13

125

Nasser, Jac 40, 75
Nissan (and models) 48

Peter Sellers Racing Team 43
Peugeot (and models) 46-49
Porsche (and models) 21, 45
Pregliasco, Mauro 41

Rallies:
    1000 Lakes/Neste/Finland 48, 75, 81, 82, 95, 97, 100, 107, 116
    Acropolis 55, 65, 66, 72, 79, 81, 97, 100, 106, 117
    Argentina 48, 49, 67, 72, 80, 81, 97, 100, 106, 107, 115
    Australia 67, 69, 70, 89, 90, 98, 100, 111
    Corsica/Tour de Corse 14, 21, 47, 55, 59, 60, 62, 64, 65, 79-81, 88, 89, 100, 104, 105
    Indonesia 95-97, 100, 108, 110, 111
    London-Mexico World Cup 43
    Manx 73, 87
    Monte Carlo 21, 28, 31, 47, 51, 53, 55, 56, 58-61, 65, 74-78, 80, 81, 87, 88, 92, 93, 95, 100-102, 104, 114, 116
    New Zealand 68-70, 80, 81, 90, 100, 107-109
    Paris – Dakar 48
    Pirelli 87
    Portugal 14, 31, 47, 58, 59, 61-63, 65, 76, 79, 89, 100, 104, 116
    RAC/Rally of GB 21, 32, 43, 56, 73, 74, 84-86, 88, 90, 91, 111, 113, 116, 118, 119
    Safari 11, 21, 78, 94, 95, 100, 102, 104, 116, 117, 119
    San Remo/Italy 71, 72, 81-84, 98, 99, 108, 119
    Scottish 52, 87
    Spain/Catalunya 47, 59, 72, 73, 90, 93, 104
    Sweden 18, 31, 33, 58, 60, 76, 89, 92, 95, 97, 100, 101, 103, 104, 118, 119
    Talavera 23, 50, 51
    Ulster 73, 87
    Welsh 87

RAS Sport team 40, 41, 74, 75, 87, 88, 90
Reisoli, Vittorio 41
Rootes Group (and models) 43

Scammell, Dick 100
Schmidt team 40
Southgate, Tony 45
Special Vehicle Engineering 23, 24
SPG 73
Subaru (and models) 16, 17, 49, 60, 69, 70, 72, 82, 86, 88, 90, 91, 93, 101, 105, 106, 108
Sutton, David 41

Taylor, Henry 43
Taylor, John 30, 36, 38, 41, 42, 45, 46, 49, 50, 64, 92, 95
TC Prototypes 23
Thompson, John 25
Toivonen, Henri 14
Toyota (and models) 9, 16, 17, 47, 49, 55, 57, 58, 60, 67, 68, 72, 73, 75, 76, 82, 86, 92, 93, 101, 115, 116
Turner, Stuart 9, 12, 32, 36, 41, 42, 45, 46, 50, 51, 118, 119
Twice Lucky 42
TWR 43

Vauxhall (and models) 43, 46
VW (and models) 21

Walkinshaw, Tom 43
Wheeler, John 9, 12, 19, 22, 25, 27, 36, 37, 41, 43, 45, 54, 55, 75
Whitaker, Martin 32, 38, 42, 97, 100, 118, 119
Wilson, Malcolm 29, 30, 32, 34, 40, 42, 46, 48, 52-54, 73, 78, 84, 86-88, 90, 91, 106, 118
World Rally Championship for Drivers 17, 47-49, 73, 75, 96, 111
World Rally Championship for Makes 17, 46, 52, 72, 87, 111

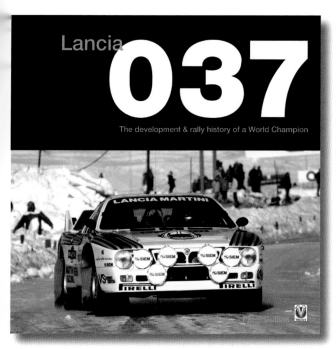

£39.99* • Hardback • 978-1-845840-76-1
• 224 pages • 300 colour & b&w photos

The world of rallying was changed forever on January 1st 1982 with the introduction of the new Group B rules. These virtually gave manufacturers carte blance to design the fastest car they could for World Rally special stages, so long as they built at least 200 identical examples. This is the story of Lancia's mid-engined and supercharged 037. Featuring many of designer and chief engineer, Ing. Sergio Limone's own photographs taken during development, it is a fascinating account of one of motorsport's most exciting and controversial vehicles.

£29.99* • Hardback • ISBN 978-1-84584-055-6
£19.99* • Paperback • ISBN 978-1-84584-063-1
• 224 pages • 480 colour pictures (both eds)

The Lancer name conjures up many images. For some, it evokes the first generation cars which fought with the best on the Safari Rally and came out the victors. Others will remember the second generation models, and who could not be aware of the Evolution (Evo) series, launched in 1992? The Lancer Evolution is not only one of the greatest rally cars of all time, it is also a desirable high-performance road car. Written in Japan with the full cooperation of Mitsubishi, this is the definitive story of all the world's Lancers, whether they carried Mitsubishi, Dodge, Colt, Plymouth, Valiant, Eagle,

*p&p extra. Call 01305 260068 for details. Prices subject to change.

£45* • ISBN 978-1-845841-09-6
• **Hardback • 256 pages • 400 mainly colour pictures**

Written with affection, appreciation and authority – and criticism where it is due – Porsche's rally story is a subject that any rally fan will find compelling. Written by lifelong Porsche enthusiast and world authority Laurence Meredith, this is a comprehensive study of Porsche's occasional foray into the world of international rallying. Illustrated with 400 photographs and with details of every rally car, it is a must for any Porsche or motorsport fanatic.

£29.99* • ISBN 978-1-845840-87-7
• **Hardback • 192 pages • 500+ illustrations**

This unique book highlights how the use of distributor-less ignition, six speed gearboxces and modern tyres all helped to bring the mighty Mini to the fore of international rallying and racing. Includes many previously unseen photos of the car's development, copies of Rover's internal documents, and pages from the road books of top rallies.

£24.99* • ISBN 978-1-904788-68-3
• **Hardback • 192 pages • 205 photos and illustrations**

The inside story of the legendary BMC Works Competitions Department as told by the three Competition Managers. Based on previously unpublished internal memos and documents, and the recollections of the prime movers, here are the ups and downs, and the politics of big time competition in an exciting era.

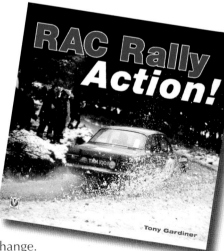

£35.99* • ISBN 978-1-903706-97-8
• **Hardback • 208 pages • 330 pictures**

This book covers the pre-WRC golden years, the Rally of the Forest period. With access to crew notes and manufacturers' archives, and containing many previously unpublished pictures, the history and excitement of the RAC International Rally of Great Britain has been captured in *RAC Rally Action!*

*p&p extra. Call 01305 260068 for details. Prices subject to change.